Author's Disclaimer
This book is not intended as medical advice. It is written
solely for informational and educational purposes. Please consult a
health professional before taking any natural modulator. Because
there is always some risk involved, the author and publisher are not
responsible for any adverse affects or consequences resulting from the
use of any of the suggestions or preparations described in this book.
The publisher does not advocate the use of any particular product, but
believes the information presented in this book should be available to
the public.

D1456159

Beta Glucan:
Nature's Secret

Second Edition

by

Vaclav Vetvicka, Ph.D.

Beta Glucan: Nature's Secret
by Vaclav Vetvicka, Ph.D.

Second Edition

Printed in the United States of America

Vaclav Vetvicka, Ph.D.
Louisville, Kentucky
email: vaclav@iglou.com

Printed on #70 Domtar Husky Opaque
Text and Headlines set in Georgia Font

Library of Congress Control Number: 2011928284
ISBN: 9780984144518

1. Beta Glucan 2. Immune System 3.Immunomodulators

TABLE OF CONTENTS

1

WHY THIS BOOK?

I wrote this book for several reasons. The first one is because I have received more and more questions about my work and particularly about beta glucans. It seemed that many people have heard about them, but are also confused about them. Despite the enormous scientific research that has gone on regarding individual glucans, very little accurate information is available on the internet. Information on the internet is often difficult for the general public to understand, or it is inaccurate (which is even more often) because it is commercially-driven. These days, the internet is obviously the first port of call for anyone looking for information. Having researched beta glucans extensively for over 20 years, I am eager to provide scientific and reliable information about them to the general public, so people can manage their health or illness more effectively.

During my 20 years of glucan studies, I have worked together with a glucan group at the Institute of Microbiology, Czech Academy of Sciences in Prague, Czech Republic. Later, I collaborated with federally-funded research on glucan activities and, together with Dr. Gordon Ross, one of the true glucan pioneers, elucidated the glucan-CR3 receptor interaction and the concept of synergy of glucan with specific antibodies. Most of today's clinical trials are based on these findings.

Later, I collaborated with several companies around the world, including the Laboratoires Goémar in Saint-Malo, where I studied the effects of marine glucans with Dr. Jean-Claude Yvin and Dr. Edouard Panak. Subsequently, I was also involved with the Groupe Roullier in Saint-Malo, doing additional research on using glucan in combination with other bioactive materials that might increase the already impressive actions of glucan. In addition, I am collaborating with the Brazilian glucan manufacturer Bioregion. Additional experiments evaluated the biological properties of glucan made and sold in the United States, Turkey, Croatia, Germany, Taiwan, South Korea, Japan and the Czech Republic. Please know that, despite all these collaborations and commercial tests, I am not selling glucan. I wrote this book objectively and completely without any profit motive.

Based on the success of my previous book *Beta Glucan: Nature's Secret,* and on feedback from the readers, I decided to write a second edition. During the time between finishing the first and second book, significant research has been done on glucan activities and effects, bringing about interesting and clinically important knowledge. It is my desire to share this knowledge with the public.

The book exists primarily to give the reader a complex overview of beta glucan's activities, potentials and setbacks. Some authors tried to write an abbreviated, highly simplified, easy-to-read, "plain English" book. I chose the opposite direction: keeping the text simple, but giving full details. Where necessary, I used scientifically common terms, which are explained later in **Chapter 19: Glossary of Terms.** On the other hand, readers seeking even more details can take advantage of numerous important publications, summarized in **Chapter 20: Scientific References.**

Since I am not in the glucan-selling business, you will not find an endorsement of any company manufacturing or selling glucan. Any commercial name used in this book is mentioned only as a result of my own extensive testing. Therefore, all information is for educational purposes only. It is my intention to give the reader enough tools to evaluate the sometimes rather bombastic statements and claims of commercial companies. After reading this book, one should have enough information (including references of the most relevant scientific studies) to clearly and safely distinguish between true data and smoke and mirrors and to pick the respectable, safe and, most of all, biologically active glucan.

I would also like to mention that there are other agents that stimulate the immune system. However, glucans are in a class apart, because those agents can push the immune system to over-stimulation. This means that they can make matters worse in the case of auto-immune illnesses such as lupus, multiple sclerosis, rheumatoid arthritis, allergies and yeast infections. Glucans, however, do stimulate the immune system, but never to the point where it becomes overactive. In addition, glucan is one of a few natural immunomodulators for which we know not only the composition, but also the mechanism of action. However, despite decades of extensive research and numerous clinical trials that are currently under way worldwide, we still do not know everything about glucan and its action. Therefore, please follow the golden rule and if in doubt, consult your physician.

And why did I write this Second Edition? There is no doubt that it would be much simpler to print another batch of the original book. However, I felt that since the time I finished writing the original volume, significant progress has been achieved in this field. I am glad to be able to inform you that none of the claims described and documented in the first edition has been found to be false. The new research has brought additional insights into the multiple roles glucan plays in influencing biological reactions. Therefore, I decided to upgrade the information and add all the significant and relevant information found in recent years.

2

INTRODUCTION

Just to avoid confusion for the readers with the many scientific words, I'll try to offer a short and, in the eyes of a real chemical expert, undoubtedly oversimplified explanation. β-D-glucans (referred to further on in this text as "glucans") belong to a group of physiologically active compounds called "biological response modifiers" and represent highly conserved structural components of cell walls in yeast, fungi, and seaweed. Generally, β-glucan is the chemical name of a polymer of β-glucose. The term "glucan" is now rather common and represents a group of chemically heterogeneous polysaccharides. There are more of these polymers and, although chemically heterogeneous, they are usually termed by the common name "β-glucans." Generally, glucans are natural polysaccharides, existing in numerous molecules of glucose bound together in several types of linkages.

In the past decades, natural glucans have sometimes been considered to be "biological immunomodulators," or "biological response modifiers" (BRMs), and sometimes as "pathogen-associated molecular patterns" (PAMPs). None of these terms are accurate, since they usually focus on only a few effects. As I will explain later, the biological effects of glucans are pleiotropic and reach a wide spectrum of biological reactions. Unlike the majority of natural products, glucans retain their full biological activity even after rigorous purification. This allows various experimental teams to evaluate and explain their biological and immunological activity on both a cellular and a molecular level.

HISTORY OF GLUCAN

Polysaccharides in general, and glucans in particular, have a long history as immunomodulators. As early as the beginning of the 18th century, it was known that certain infectious diseases showed a therapeutic effect on malignant processes. The dedicated use of such therapy dates from around the middle of the 19th century, at which time Bush performed experiments in search of curing sarcoma by infecting patients with an acute streptococcus bacterial infection of the dermis. Coley repeated these therapeutic procedures towards the end of the 19th century. However, the early researchers had no knowledge of the molecule responsible for the observed effects.

It is likely that the first investigated substance with immunomodulating properties was the so-called "endotoxin or lipopolysaccharide" (LPS) of Gram-negative microbes. A paper describing the endotoxin was published in 1865. LPS induced stronger phagocytosis with a potential protective effect for a host. However, its toxic effects were totally dominant. It was later found that a saccharidic moiety of LPS (with prevailing glucose, galactose and mannose content) is non-toxic but bears immunomodulating activity. It was apparent that polysaccharides could act as immunomodulators, even though their toxicity was negligible.

The finding mentioned above was probably the catalyst for research of other polysaccharide preparations. Further documented history of polysaccharides as immunomodulators goes back to the 40s of the last century when Shear and co-workers (Shear et al., 1943) described a substance from *Serratia marcescens* cultures that caused tumor necrosis. Subsequently, this substance (known as "Shear's polysaccharide") was identified as a mixture of three polysaccharides with the main chain consisting of D-glucose and D-mannose units connected by (1→3) glycosidic linkages.

Generally speaking, immuno-modulatory preparations from bacteria—either extracts from intact cells or isolated products such as Shear's polysaccharide— are à priori suspicious and their use can be dangerous. Perhaps due to this fact, focus was given to polysaccharides isolated from "human friendly" organisms, i.e., yeasts and edible mushrooms. Eventually, other polysaccharide immunomodulators were sought. Their investigation began in the 60s and 70s of the last century. Two lines can be traced in the history of glucan, each based on different starting points but gradually converging. The foremost

5

origins were in the United States, Europe, and Japan, respectively. Research on glucans in the Euro-American milieu was based on knowledge of the immunomodulatory effects of zymosan—a mixture of polysaccharides isolated from the cell walls of the well-known and widely-used baker's yeast *Saccharomyces cerevisiae.*

Although zymosan was able to stimulate a non-specific immune response, initially it was not clear what component of this rather crude composition was responsible for the activity. When zymosan was examined in detail, glucan was identified as the component of primary effect. It was subsequently isolated, and its immunological effects were investigated. This research was pioneered by Nicholas R. DiLuzio (DiLuzio and Riggi, 1970) of Tulane University in New Orleans. In a number of papers, he demonstrated that glucan administration caused significant phagocytic stimulation of the reticuloendothelial system, enhanced host defense mechanisms, and resistance to experimental tumors.

Intensive research of immuno-modulating activities of β-glucan was also conducted in Japan, and they arrived at β-glucan via a different route. In Asian medicine, consuming different medicinal mushrooms (shiitake, maitake, reishi etc.) has been a long tradition. In detailed studies of the biological effects of these mushrooms, in particular their anticancer action, β-glucans were again found to be the main cause of non-specific immunomodulation. This initial investigation was conducted by Goro Chihara at Teikyo University in Kawasaki, who isolated β-glucan from the shiitake mushroom, which he referred to as "lentinan," (*Lentinus edodes*, now *Lentinula edodes* {Chihara et al. 1969}). This glucan, with some subsequent modification, was later approved as a drug and has been successfully used for almost 30 years.

The important quality of polysaccharidic immunomodulators, or glucans, is evidenced by the fact that all the sufficiently purified ones distinguish themselves by very low toxicity (e.g., for mouse lentinan has LD_{50} > 1,600 mg/kg). Conversely, the considerable heterogeneity of all natural glucans continues to be the cause of a series of mutually contradicting conclusions. Therefore, attempts are sometimes made to solve this problem using semisynthetic and synthetic versions, which are ideal for accurate immunological research. Unfortunately, the costs of chemical synthesis of glucan-based oligosaccharides are too high and therefore make the commercial or clinical use of such samples impossible. Thus, the synthetic oligosaccharides, despite their high biological activities, remain outside the reach of commercial use. We can hope that, in the future, new methods of synthesis might make the use of small synthetic oligosaccharides commercially feasible.

WHY GLUCANS?

Serious diseases such as AIDS and cancer continue to be a major health concern. New flu strains are emerging faster than the Centers for Disease Control and Prevention (CDC) can document. And the traditional flu vaccines are becoming less and less effective. Clearly, despite all efforts and resources, science and medicine cannot keep up with ever emerging and transforming infectious microorganisms. Antibiotics can serve as an example: their overuse created more virulent and more resistant strains of bacteria, making infections acquired in hospitals one of the leading causes of death. Approximately 18, 000 Americans die annually from infections with MRSA (Methicillin-resistant *Staphylococcus aureus*) alone. Allergies, cancers, infections, or autoimmune diseases are the result of dysfunctional immune systems. At the same time, however, it is clear that the speed of our healing is dependent on a fully functional immunity. Natural remedies and immunostimulators are therefore more important than ever.

Natural products, useful in preventing and/or treating various diseases, have been sought after throughout the history of mankind. Some people believe that old remedies are just better than synthetic product of the pharmaceutical industry and some prefer cheaper natural products versus the expensive drugs with potential negative side effects. However, even natural products are not completely without problems. A main problem in characterizing natural products also occurs with glucans: in nature, they represent a complex mixture of ingredients, each of which might (and probably will) contribute to biological activity. Therefore, the evaluation of glucan properties had to focus not only on biochemical characteristics and biological activities but primarily on adequate isolation techniques which, in the end, gave us the material consisting of highly purified glucan molecules. As a result, the only meaningful data has come from experiments bases on sufficiently purified glucans (Vetvicka, 2001, Vetvicka and Vetvickova, 2010). Glucans have been found to provide a remarkable range of health benefits and are particularly important against the two most common conventional causes of death in industrialized countries—cardiovascular diseases and cancer. We will approach this later.

3

Individual Types of Glucan

During decades of research, numerous types of glucan have been isolated and described. In scientific literature, you can find dozens, if not hundreds, of different components, all under the name glucan. Unfortunately, not all glucans are created equal. Glucans widely differ not only in physicochemical properties such as branching or molecular weight, but also in biological properties. Unfortunately, some of the described glucans have no biological activities at all. The isolation and purification processes need to constantly monitor all conditions; otherwise, the final product will have limited biological activity, if any at all.

The various glucans are isolated mostly from yeast, mushrooms, grain and seaweed. Despite the long-term "fights" among individual groups of glucan researchers, it is clear that the preferred sources of individual glucans are based more on the traditions and the availability in individual countries than on any scientific proof of superiority.

Yeasts are the major source of glucans in Western countries. The Far East (Japan, China and East Russia) traditionally focuses on mushrooms (based on their folk remedy). A high amount of seaweed existing in France resulted in seaweed-derived glucan called "Phycarine" produced by Goemar Laboratories. Phycarine can serve as another example of how the availability of resources influenced the material used for isolation of glucan. Goemar Laboratories are located in Brittany on the western coast of France where there is an abundance of seaweed. Goemar Laboratories therefore decided to use the easily available natural resources and produced an excellent, high quality glucan (Vetvicka and Yvin, 2004). In some cases, the content of carbohydrates, which are mainly present as polysaccharides or glycoproteins, can reach up to 90%.

The commercial importance of fungal polysaccharides has drawn a great deal of attention to the field of biological response modifiers as well as functional food. In this regard, the most commonly cultivated mushroom is of the genus Pleurotus. The annual production of these mushrooms is more than 900,000 tons. A variety of genus Pleurotus species has pharmacological properties. The polysaccharides from these mushrooms are usually found under the name "Pleuran" (Zeman et al., 2001).

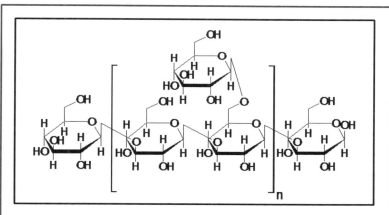

Schematic layout of glucan molecule

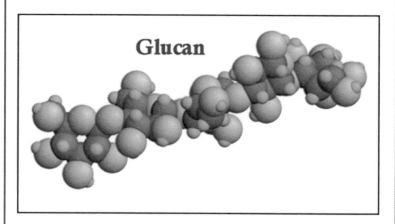

Three-dimentional shape of glucan molecule

Figure 1a-1b

Glucan derived from baker's yeast (*Saccharomyces cerevisiae*) has been the most extensively studied and has produced the highest biological effects. We recommend that readers interested in the introduction of a significant number of biologically active glucans read an excellent review written by Harada and Ohno (2008). The general structure of β-glucan is summarized in Figure 1 a. Below it is a three-dimensional picture of the same molecule (Figure 1 b).

The considerable heterogeneity of all natural β-glucans, not only from saccharomycetes but also from other sources, obviously was and continues to be the cause of a series of mutually contradicting conclusions. Research on the cell wall of different "fungal" species has not led to a straightforward model of its structure and the concepts of its organization have undergone certain development. According to Stratford (Stratford, 1994), the yeast cell wall resembles reinforced concrete. An armature, representing about 35% of the wall mass and formed by fibrils of alkali insoluble β(1→3)-glucan, is dipped into mannoproteins (about 25 - 35 % of the wall mass) and bound to the armature through amorphous β-glucan and chitin. The cell wall of fungi is schematically shown in Figure 2. The cell wall of other "fungi" is constructed in a similar manner. An excellent review of the chemistry of yeast and fungal cell wall can be found in (Kogan, 2000).

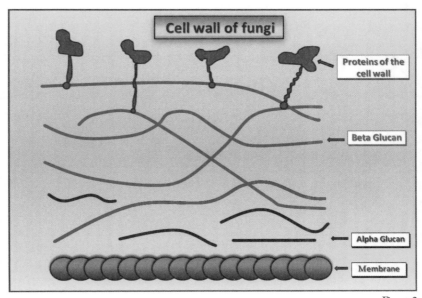

Figure 2

Glucans can be isolated from almost every species of yeast. Glucan forms part of the yeast cell wall, along with mannan, proteins, lipids and small amounts of chitin. Glucans represent a major structural component of the cell wall in fungi and some plants. Different physicochemical parameters, such as solubility, primary structure, molecular weight, branching and polymer charge, all play a role in determining whether the polysaccharide modulates immune reactions. Some conclusions can be made, though. Branched or linear 1,4 - β-glucans have very limited activity, if any. Glucans with 1,6 configuration, usually have limited activity. The highest stimulation of defense reactions has been achieved with β-glucans that have a 1,3 configuration with additional branching at the position o-6 of the 1-3 linked D-glucose residues. Among all glucans, those with a 1,3 configuration are best characterized. An excellent review of glucans as biological response modifiers and the relationship between structure and functional activity is given in (Bohn and BeMiller, 1995). The characteristics of some commercially common glucans are given in the following Table:

SOME B-GLUCANS WITH IMMUNOMODULATORY EFFECTS

Name	Source	Character of polymer
Curdlan	*Alcaligenes faecalis*	linear
Laminaran	*Laminaria sp.*	linear
Pachymaran	*Poria cocos*	linear
Lentinan	*Lentinus edodes*	branched
Pleuran	*Pleurotus ostreatus*	branched
Schizophyllan	*Schizophyllum commune*	branched
Sclerotinan (SSG)	*Sclerotinia sclerotiorum*	branched
Scleroglucan	*Sclerotium glucanicum,*	branched
T-4-N, T-5-N	*Dictyophora indusiata*	branched
Yeast glucan	*Saccharomyces cerevisiae*	branched

Until recently, biologically efficient glucans were supposed to have similar structures—the main chain of $\beta(1{\rightarrow}3)$ bound D-glucopyranose moieties to which some D-glucopyranoses are randomly connected by $\beta(1{\rightarrow}6)$ linkages. Basically, the 1,3 chain forms a glucan backbone, with additional glucose molecules branching out in 1,6 linkages. However, the detailed structure of glucans from dissimilar sources differs as well as their biological activity. In native glucans, their fibrils are composed from organized parts in which the main chain is coiled to a triple helix. The triple helix, formed by three hydrogen bonds and stabilized by side chains, is most likely present only in high-molecular glucans with molecular weight over 90 kDa. Conversely, in isolates of glucans, this triple-helical structure can be killed during an isolation process; for the H-bonds of triple helices are interrupted by increased temperature, high pH or certain solvents. The next table shows the degree of branching in various commercial glucans.

DEGREE OF BRANCHING OF DIFFERENT β-GLUCANS

β-glucan	Source	Degree of branching
Pachymaran	*Poria cocos*	0,015 - 0,02
Yeast glucan	*Saccharomyces cerevisiae*	0,03 - 0,2
Lentinan	*Lentinula edodes*	0,23 - 0,33
Pleuran	*Pleurotus ostreatus*	0,25
Grifolan	*Grifola frondosa*	0,31 - 0,36
Schizophyllan	*Schizophyllum commune*	0,33
SSG	*Sclerotinia sclerotiorum*	0,5

Since glucans occur in diverse locations in the fungal walls, the methods of isolation greatly depend on their type and occurrence. If the polysaccharide is a part of an extramural capsule, it can be rather easily isolated by washing the cells and the subsequent precipitation. Isolation of water-insoluble glucans involves extraction at an elevated temperature and subsequent separation of polysaccharides and other cellular components by appropriate methods.

Since it was determined that the lethal dose of glucan is extremely high, glucan was found to be worthy of consideration in clinical practice. After detailed clinical trials, the Japanese government approved the use of glucan for the treatment of stomach and colorectal cancer, particularly when used together with cytostatic chemotherapy drugs such as mitomycin C and 5-fluorouracil.

The original studies on the effects glucan has on the immune system focused on mice. Subsequent studies demonstrated that glucan has a strong immunostimulating activity in a wide variety of other species, including earthworms, shrimp, fish, chicken, rats, rabbits, guinea pigs, sheep, pigs, cattle, and humans. Based on these results, it has been concluded that glucans represent a type of immunostimulant that is active over the broadest spectrum of biological species and that it is one of the first immunostimulants active across the evolutionary spectrum. Some experiments even show that glucan can help in the protection of plants. Glucan is therefore, not only a biologically active polysaccharide with strong immunomodulating effects, but is also considered to be an evolutionary and extremely old stimulant of a variety of defense reactions (Vetvicka and Sima, 2004).

Despite extensive investigations, no consensus on the source, size, or other properties of glucan has been reached. An important comparison of yeast-derived and mushroom-derived glucans, and their biological activities, is given in (Kogan, 2000, Vetvicka and Vetvickova, 2007 a, b, 2010). One clear disadvantage of mushroom glucans is a common and usually strong smell that can be unpleasant to some.

In addition, numerous concentrations and routes of administration have been tested. These include intraperitoneal, subcutaneous, and intravenous applications. For many years, oral treatment with glucan has been on the periphery of interest, despite the fact that it represents the most convenient route. The reasons were the lack of knowledge about the possible route of glucan through the gastrointestinal track and about the interaction with cells in the Payers patches and in the gut. However, in the last decade a renewed interest in human application has brought about some important studies of orally-administered glucan.

4

DIFFERENCES AMONG INDIVIDUAL GLUCANS

Though various models of the fungal cell wall differ somewhat, they concur in that β-glucan is not located on the surface of the wall but is more or less immersed in the wall material. With regard to both immunological research and pharmaceutical utilization of glucans, an important conclusion can be reached. In macroorganisms, glucans initially act as markers of fungal invasion, allowing the activity of β-glucan preparations to increase with the degree of removal of glucan fibrils.

There are various natural sources of β-glucans; however, they are most frequently prepared from fungal cell walls. Baker's yeast is the most common and likely the best raw material for glucan extraction. It is rather difficult to get the glucan molecule out from the yeast wall structure in a highly active form. The major challenge is to remove the impurities, such as manno-proteins and lipids (attached to the end points of the side branches in the intact cell wall), without the loss of any biological activity. On the other hand, whole yeasts alone are not the optimal source of active beta glucan, so we cannot begin to consume whole yeast and expect to observe the same effects as with glucan supplement. This is due primarily to their content of available glucan not being high enough, which may result in impurities acting against the biological effects of glucan molecules. In addition, due to their large size, they are normally not phagocytosed by the gut cells and, therefore, insufficient potential glucan can enter our body.

Until recently, biologically efficient β-glucans were supposed to have similar structures—a main chain of β(1→3) bound D-glucopyranose molecules to which some D-glucopyranoses are

14

randomly connected by β(1→6) linkages causing a different degree of branching in different glucans. However, the detailed structure of β-glucans from dissimilar sources differs as well as their biological activity. In native β-glucans, their fibrils are composed from organized parts in which the main chain is coiled to a triple helix. These regions are combined with single or double filaments of β(1→3)-D-glucopyranoses. The triple helix, formed by three hydrogen bonds and stabilized by side chains, is probably present only in high-molecular β-glucans with molecular weight over 90 kDa (Ohno et al., 1988). The hydrogen bonds of triple helices can be interrupted by increased temperature, high pH or certain solvents.

Diverse data on comparison of structure, molecular size, and biological effects can be found in the literature. For example, the antitumor activity of schizophyllan is supposedly conditioned by the triple helix presence and a molecular weight higher than 100 kDa. It is more than likely that the triple helix structure is not the sole effective form of β-glucan, because alkalic treatment, used in most isolation procedures, kills this structure. In addition, the most recent opinions do not confirm the established ideas of the necessity of high molecular mass and branching of biologically active β-glucans. Thirty-three years ago, Kabat (Kabat, 1976) found that, for antigen polysaccharidic determinants, the size of the binding site on an antibody corresponds to six or seven monosaccharide units. The size of the binding site for β-glucan (in this case on a receptor of an immunocompetent cell, e.g., the macrophage) also appears to correspond to this number of glucose residues (Kabat, 1976). These findings were recently used in preparation of small, synthetic glucan-based oligosaccharides (Descroix et al., 2010).

From the information mentioned in this chapter, it is clear that our knowledge of glucan chemistry is still far from complete. The same is true about the correlation of size, origin, structure, and biological activities. Clearly, the older assumption that only large, insoluble glucans are biologically active was wrong. This is most probably due to the fact that glucans from resources such as seaweed or grains were insufficiently purified, resulting in mediocre biological activities. To summarize, information about the molecular size, branching, solubility, three-dimensional structure etc. is interesting, but is most probably not relevant for the real-world situation. More important are the data about purification, isolation, characterization, and final purity of the sample.

5

BASIC PRINCIPLES
OF THE IMMUNE SYSTEM

IMMUNITY: WHAT IS IT ALL ABOUT?

Every organism is essentially a miracle of harmony, with individual parts—cells, tissues and organs—perfectly dividing all necessary functions. This beautiful system can, however, be easily disturbed by anything that endangers the integrity of our body. The defense systems (and in reality, the immune system is only one of several defense systems) were developed to protect the organism against these threats. Generally, organisms have developed various defensive mechanisms that help them to survive - the ability to run away from danger, grow large teeth to fight, or the ability to hide. Additional defensive tools are external or internal mechanical barriers (such as skin or mucosal membranes).

Immunology as a science describing the immune system, was born as a part of the microbiological sciences. The concept of immunology was predominant among both laymen and specialists, until at least the first half of last century. However, it was only during the past 40-to-50 years that immunology established itself as an independent, rapidly developing, and challenging science. A broad array of scientists, a number of them Nobel Prize winners, laid the foundation for modern immunology. The term "immunity" derives from the Latin *immunis*, meaning "free of burden". A person resistant to a certain disease is therefore said to be "immune" to it. An ever-increasing mass of evidence has since then confirmed that defense (immune) mechanisms are, generally speaking, an integral part of homeostasis, which is indispensable for the maintenance of an individual's integrity.

Three major principles of immunity are common to all living creatures. These principles are recognition, processing, and response (usually elimination). Generally speaking, all creatures stay healthy by employing these "simple" principles. It is necessary to *recognize* invading danger by distinguishing between self (i.e., everything constituting an integral part of an organism) and non-self (all the rest). Recognition is followed by *processing of* this information (this analysis determines the precise events of the response) and by *elimination* of the threat (either by specific antibodies or by the action of various cell types). For millions of years, each of these mechanisms has been endowed with a memory.

In general, the immune system is a highly complex system of cells, organs and biological molecules. However, host defenses do not work alone. They work in close collaboration with other physiological systems including the central nervous system and the endocrine system.

The Immune System – The Key to Staying Healthy

The immune system has evolved during millions and millions of years as an extremely potent and efficient defense mechanism directed toward a single goal: to keep us healthy. The vertebrate immune system has evolved to detect and respond to the vast array of microbes which it encounters on an ongoing basis. The innate immune system is the first line of host defense and its underlying mechanisms can occur within minutes of encountering a potentially threatening microbe, which results in a variety of responses, such as the production of cytokines and chemokines, and the presentation of microbial antigens to lymphocytes. These reactions trigger the adaptive arm of the immune system. Our immune system consists of a vast and intricate network of various cell types that constantly monitor our health. A simplified scheme of immune reaction is given in Figure 3. However, despite numerous factors involved—including macrophages, several types of lymphocytes, antibodies and other humoral factors—the ever decaying conditions we live in are constantly lowering our immune responses and decreasing the ability of our body to survive constant attacks.

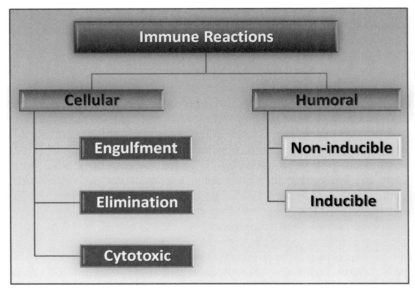

```
                    Immune Reactions


       Cellular                      Humoral


          Engulfment                    Non-inducible


          Elimination                   Inducible


          Cytotoxic
```

Figure 3

During the last century, the stress our defense systems are working under has significantly increased. The various pressures in our lives, such as psychological stress, dangerous chemicals in the form of pollution, car exhausts, allergens, etc. have multiplied and led to ever decreasing levels of our immune responses. The immune system is further weakened by age and by diseases we may contract. Severe allergies to the everyday environment were virtually unknown as recently as a hundred years ago. Today, one out of two people suffers from some type of allergenic problem. To compound the situation, our environment is deteriorating more each year. And, as a result, our immune system is becoming less effective. In addition, repeated exposure to toxic chemicals such as pesticides and the constant attacks of pathogens, further diminish the ability of our body to fight disease. Fifty years ago, during the rise of antibiotics, the situation seemed to be under control. It appeared that every disease could be combated with a specific and highly potent treatment. During the ensuing decades, antibiotics were grossly overused. This resulted in significant resistance to them by the most common bacteria. Additionally, conditions such as increased environmental and psychological stress, as well as the evolution of cancer, continue at an unprecedented rate. Despite great achievements and decades of intensive, labor-consuming and expensive research, the incidence of various tumors and cancers continues to increase at an alarming rate. Based upon the National Cancer Institute estimates, 1-in-5 humans in the U. S. are likely to be a victim of cancer in their lifetime. This is considered to be of an epidemic proportion.

How our Immune System Works

To better understand how glucan really works, one must first understand a little about the immune system. The immune system is a system of cells, organs, and soluble molecules working in unison to defend the body against foreign pathogens. This system consists of numerous components, constantly on alert to find invading pathogens, finding means to destroy them, and eliminating them from our body. Individual cells interact with one another using physical contact and/or the secretion of various bioactive molecules.

Generally speaking, scientists recognize two basic types of immune system: the _innate_ immune system and the _acquired_ immune system. The innate system (sometimes called "non-specific") is considered to be the first line of defense and represents a significant part of the entire immune system. It includes mechanical barriers, cells such as macrophages and neutrophils and soluble factors such as the complement system and antimicrobial peptides. In many, if not most, cases of infection, the innate immune mechanisms are sufficient to prevent full-blown infection.

The acquired immune system (sometimes called "specific immunity") identifies the characteristic proteins of invading microorganisms and their toxins. This part of immunity consists of cells such as T and B lymphocytes and antigen-presenting cells (macrophages and dendritic cells). The lymphocytes recognize the invading pathogens by specific antibodies (B lymphocytes) or specific receptors (T lymphocytes). Microbial adversaries have tremendous ability through mutations to evolve evasive strategies, and it is clear that a very large number of specific immune defenses need to be at the body's disposal. Antibodies represent a perfect solution. They are highly adaptable molecules that can recognize and stick to the offending microbe, activate the complement system and stimulate the macrophage/dendritic cell system of antigen presenting cells. Because our body can make hundreds of thousands and probably millions of different antibody molecules, it is not feasible to have a large number of lymphocytes producing each type of antibody. Upon recognizing an antigen, some lymphocytes undergo successive waves of proliferation and build up a large clone of specific antibody-making cells. In general, our body produces five classes of antibodies (often called immunoglobulins): IgM (approx. 8 %), IgG (76 %), IgA (15 %), IgD (1 %), and IgE (0.002 %). IgM antibodies represent a first response, as they are formed first and are the most multipurpose (we call this response a "primary" one). The other types of antibodies are formed later and therefore give way to what is termed a "secondary antibody response".

The antibodies involved in immune reactions are made by cells called "B lymphocytes." These cells are responsible for the production and secretion of highly specific antibodies that subsequently recognize bacteria and guide macrophages and T lymphocytes towards objects coated with them. The antibodies act as a type of targeting device that aims directly at the membrane of cells and microorganisms from (remove "from"), which our body is trying to eliminate. When macrophages and lymphocytes are insufficient, glucan comes to the rescue. Numerous experiments clearly demonstrated that when glucan is used simultaneously with anti-cancer antibodies, the healing is much stronger than only using glucan or antibodies alone (Hong et al., 2003, 2004).

The acquired immune system is further supported by the existence of memory. Upon the successful fight with an infection, a very small number of lymphocytes that remember the first contact with bacteria are left behind. In case of any subsequent exposure, these cells are responsible for a faster and stronger immune reaction. The existence of memory cells is the basis for vaccination, using a harmless form of the infective agent for the initial injection.

Another class of lymphocytes, the T lymphocytes, does not make antibodies but is focused on control of intracellular infections. These cells originate in bone marrow and are educated in the thymus. Similar to the B lymphocytes, each T lymphocyte has its individual receptors that recognize antigens, and upon binding, undergo a clonal expansion resulting in effector and memory cells. In addition, we recognize cytotoxic T lymphocytes, which release cytokines causing cell death and/or attracting macrophages and other cells to the area. An additional type of T lymphocytes includes the so-called suppressor

20

T cells, which negatively regulates the immune response, thus serving in the role of a biological pressure valve. Finally, the best known fractions are the so called "T helper" cells (often known as CD4-positive cells), which directly or indirectly regulate the immune response of other immunocytes.

Generally, there are three different types of cells involved in immune reactions: the "T" and the "B" lymphocytes (see above), and the last major cell type, the macrophage. The immune system's first line of defense is formed by macrophages, and they are often considered to be the most important players of all the cells involved in defense reactions. These cells are present in every single organ of our body, including peripheral blood and various organs such as the spleen, lymph nodes, liver, skin and brain. Macrophages constantly monitor their surroundings for anything they consider to be foreign. These cells control circulating streams of body fluids (such as blood and lymph) and react adequately to any observed changes. Macrophages are present in every single tissue of our body. And no significant change can occur without being spotted by these warriors.

The macrophages are final, mature types of cells derived from pluripotent stem cells localized in the bone marrow. The process of cell differentiation within the myeloid lineage requires several steps. The most immature cells are called the "monoblasts" that further divide into "promonocytes". The promonocytes become monocytes that reside in bone marrow for some period and mature under the instructive treatment of local microenvironment (where glucan can play an important role). Following this, the monocytes leave the bone marrow by a random process and through blood circulation, migrate into various tissues and organs of the body. There they undergo further differentiation and maturation until reaching their final stages of development. It is important to note that the cells of the granulocyte (neutrophil) series possess similar properties. Their origin is very close to that of the myeloid lineage. Readers seeking more information about macrophages and their role in immunity can read a comprehensive review (Fornusek and Vetvicka, 1992).

Invading microorganisms, mutated or damaged cells, and even macroscopic material such as small parts of wood or dust, are considered foreign and potentially harmful. The importance of macrophages can be further stressed by the fact that these cells are the evolutionary oldest and best-preserved defensive cells. There is not a living creature on Earth that does not depend on macrophages for its immune reactions.

In order to fulfill their defensive duties, macrophages have to change from a steady-state to an activated state. Only after becoming fully activated can macrophages perform their numerous functions. After several steps in the activation cascade, an entire sequence of complicated, and highly sophisticated, metabolic changes occurs, which results in changes on the membrane of the macrophages.

The basic function of macrophages is to analyze the situation, recognize the invader, and phagocytose their prey. Phagocytosis is the most pronounced function of macrophages, and it is phylogenetically the oldest type of defense mechanism in the animal kingdom. After the contact with its prey, the internalized material is killed by potent intracellular enzymes. However, macrophages can do even more. Macrophages are able to produce, and actively secrete, numerous factors in order to destroy attacking bacteria and significant parts of pathogens. An activated macrophage represents a veritable powerhouse. Macrophages can switch between active and resting states, so there is no fear of the immune system being over stimulated by taking glucan supplements. Figure 4 shows the phagocytosis of bacteria (Figure 4 a) or phagocytosis of synthetic microspheres often used for glucan evaluation (Figure 4 b).

The key role that macrophages play in host defense has gradually become understood through years of research. The truth is that we can find macrophages in all multi-cellular organisms that have been investigated thus far and often in creatures in which no other immune cells exist. Besides their role in immunity, macrophages play a crucial role in processes such as tissue turnover, tissue remodeling during embryogenesis, tissue destruction and repair during injury, and tissue renewal. Their secretory function is nonsubstitutional; over a hundred distinct molecular entities have been found to be secreted by macrophages. Additionally, the spectrum of microorganisms kept in check by macrophages (or even more generally by phagocytes) includes fungi, bacteria, and virus-infected cells. It can be said that macrophages are one of the most versatile cells in multicellular organisms; and their role in the masterminding of immune reactions is still far from being completely understood.

The cellular branch of immune reactions has another ace up

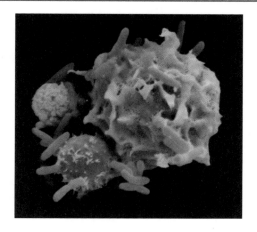

Macrophage phagocytizing bacteria (green).

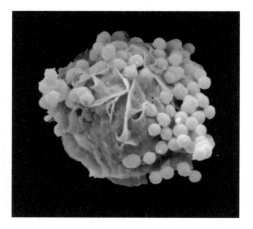

Macrophage phagocytizing synthetic HEMA particles(blue).

Figure 4a-4b

its sleeve, the T lymphocytes. These small cells function in close cooperation with macrophages, and upon activation, seek and kill, not only viruses, but also cancer cells. Lymphocytes activated via the macrophage network, or some specific drugs, can become what is called "natural killer cells (NK cells). These NK cells have one purpose only: to recognize and kill dangerous cells that result from either viral infections or carcinogenic mutations. They are essentially a bloodthirsty subtype of killer lymphocytes. Their potential to kill invaders, or damaged/malignant self cells, is really substantial.

All types of immune cells work together in harmony. They constantly interact with each other and function on the basis of information transferred using a complicated network of humoral factors such as enzymes and cytokines.

POSSIBLE PROBLEMS

Every infectious attack lowers the level of the body's natural immunity. However, infection is not the only problem we must address. Today, we are being assaulted by a number of negative influences that contribute to the impairment of defense mechanisms. Among these factors are polluted environment, persistent stress, poor nutrition or lifestyle, various types of radiation, and the overuse of antibiotics. These are some of the catalysts that may lead to immune system deficiencies that are manifested by problems such as: overall fatigue, slow-healing wounds, repeated infections, lowered numbers of blood cells, and increased susceptibility to various oncological diseases. Clearly, our body needs all the help we can get. Therefore, nutritional supplements supporting and activating individual parts of the defense reactions are becoming extremely critical.

Why do we become ill if we have an immune system? Our immune system is doing all it can, but we have to remember that we live under dangerous conditions, with millions of bacteria, viruses and parasites constantly entering our bodies. Considering the danger, most of us are usually fairly healthy. However, we become ill because our immune system somehow has been weakened. It can no longer do its job, or at the least not as well. This can happen for different reasons: inadequate diet, excessive alcohol consumption, stress, lack of sleep, poor intake of vitamins and minerals, excessive intake of medicines, lack of exercise, and pollution, to name only a few. As a result, our immune system needs all the help it can get. In the case of cancer, the fact that cancer cells are very clever, and can disguise themselves as "self" cells, makes them more difficult for immune cells to identify and kill. This is why an immunomodulator like glucan is helpful.

6

MECHANISMS OF GLUCAN ACTION

Natural products useful in preventing or treating diseases have been highly sought after throughout the history of humanity. A major problem in characterizing many natural products is that they represent a complex mixture of ingredients, each one of which may contribute to their bioactivity. Glucans from fungi, yeast, and seaweed are well-known biologic response modifiers that function as immunostimulants against infectious diseases and cancer. Unlike most other natural products, properly purified glucans retain their bioactivity. This allows us to characterize how glucans work on a cellular and molecular level.

Glucan has been used as an immunoadjuvant therapy for cancer since 1980, primarily in Japan (Takeshita et al., 1991). Another activity demonstrated with β-glucan in the mid 1980's was its ability to stimulate hematopoiesis in an analogous manner as a granulocyte-monocyte colony-stimulating-factor (Patchen et al., 1986). Both particulate and soluble β-glucans, administered intravenously, caused the significantly enhanced recovery of blood cell counts after gamma radiation. Repeated studies confirmed that glucan could reverse the myelosuppression produced from chemotherapeutic drugs.

In addition to their effect in the treatment of cancer, glucans have been demonstrated to protect against infection with both bacteria and protozoa in several experimental models and were shown to enhance antibiotic efficacy in infections with antibiotic-resistant bacteria. The protective effects of glucans were shown in experimental infection with *Leishmania major, Leishmania donovani, Candida albicans, Toxoplasma gondii, Streptococcus suis, Plasmodium berghei, Staphylococcus aureus, Escherichia coli, Mesocestoides corti, Trypanosoma cruzi, Eimeria vermiformis and anthrax (Bacillus anthracis).*

Despite long-term interest and research, the mechanism of how glucan affected our health remained a mystery. During the last 30 years, predominantly in the Japanese pharmaceutical literature, over 500 scientific papers have examined glucan structure but only in relation to tumoricidal or bactericidal activity. These papers did not attempt to identify its target receptors in order to define optimal polysaccharide structure. Overall, these studies confirmed that glucans, either soluble or particulate, exhibit strong biological effects.

Only in the last decade, extensive research by numerous scientific groups has helped to reveal the extraordinary effects that glucan has on our immune system. Experiments done by a University of Louisville research group, led by professor Gordon Ross, focused on one particular receptor called complement receptor type 3 (CR3 receptor) as a promising target of glucan. Subsequent detailed analysis of the interaction of human cells with glucans has demonstrated that the CR3 receptor is primarily responsible for both the binding and biological effects of glucans. CR3 is considered to be the most important receptor mediating clearance of opsonized immune complexes by the phagocytic system (Ross and Vetvicka, 1993).

In addition to its function as a receptor for cytotoxicity and phagocytosis, it also serves as an adhesion molecule responsible for leukocyte diapedesis. For these functions, the CR3 molecule goes through a series of inside-out and outside-in signaling steps resulting in exposure of high-affinity binding sites. In 1987, it was shown that neutrophil CR3-dependent phagocytosis and degranulation in response to iC3b-opsonized particles required ligation of two different binding sites in CR3—one for iC3b and one for beta-glucan. Using fluorescent-labeled glucan and CHO cells expressing recombinant chimeras, the binding site was mapped to a region of CR3 located C-terminal to the I-domain. This information revealed the mechanisms of glucan action. The activation of CR3 by glucan is initiated through the binding of glucan to a lectin site in the CR3 molecule (see Figure 5). The CR3 molecule is formed by a two-chain structure with several biologically important domains, each binding different molecules.

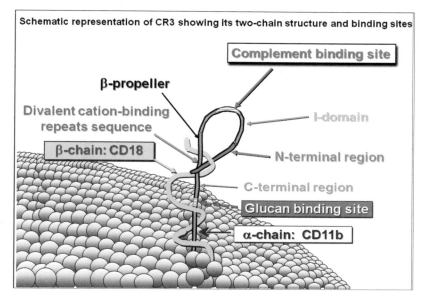

Schematic representation of CR3 showing its two-chain structure and binding sites

Complement binding site

β-propeller

Divalent cation-binding
repeats sequence

I-domain

β-chain: CD18

N-terminal region

C-terminal region

Glucan binding site

α-chain: CD11b

Figure 5

After the binding, the CR3 is primed for cytotoxic degranulation in response to the binding of iC3b fragment to a different part of the CR3 molecule. Detailed studies later showed that soluble β-glucan binding to the lectin site of neutrophil or natural killer cell CR3 generates a primed state of that receptor capable of mediating cytotoxicity of iC3b-opsonized target cells (Vetvicka et al., 1996). These data were further validated by the use of cells from CR3-deficient mice that were resistant to the effects of glucan. Similar to the situation with leukocytes, CR3 that is present on natural killer cells (NK cells) functions in a like manner. Most of these studies were done in cancer models. However a similar mechanism applies to microbial pathogens. For more details about how the CR3 receptor works and how this hypothesis was confirmed by the use of CR3-deficient mice, see these excellent papers (Thornton et al., 1996, Xia and Ross, 1999).

Research in the past decade has clearly shown that glucans specifically target macrophages, NK cells and neutrophils to tumor cells that are opsonized with complement fragments and antibodies, which in reality gives glucan the same specificity as the tumor-opsonizing antibodies (Ross et al., 1999). To summarize, our bodies need two things to occur simultaneously: microbes/tumor cells need to be coated by complement fragments and glucan has to be present. The optimal immune response takes place only when both events occur.

ADDITIONAL RECEPTORS

To successfully exploit the biological effects of these carbohydrates, and improve host defenses against fungal pathogens, it is important to continue investigating the receptors involved in β-glucan recognition. A number of cells have been shown to express β-glucan receptors, since they were initially identified on monocytes. These comprise both immune and nonimmune cells, including neutrophils, eosinophils, natural killer cells, dendritic cells, endothelial cells, alveolar epithelial cells, fibroblasts, and a variety of macrophages including microglia. As mentioned earlier, there are now a number of specific β-glucan receptors identified.

In addition, glucan can influence immune cells by binding to other receptors. Toll-like receptors were not discovered until quite recently, although they possibly represent the most important receptor molecules of the non-adaptive component of the immune system. The name of these receptors is derived from sequential homology with a protein coded by the Toll gene. This gene occurs in *Drosophilla* flies, where it helps in defense against fungal infection. Thus far, approximately eleven toll-like receptors are known. Glucan (and also zymosan, intact yeast cells, LPS) is initially bound to toll-like receptor 2.

Dectin-1 is a lectin located on the macrophage surface with a transmembrane stalk region and cytoplasmic tail and represents a Group V member of the C-type lectin superfamily. This receptor has a special application in the detection and phagocytosis of fungal pathogens. In certain cases, it cooperates with the toll-like receptor 2. It is also a transmembrane protein with many particular functions, e.g., binding of a fungal PAMP, uptake and killing of invading cells, and induction of the production of cytokines and chemokines. The binding of glucans and zymosan or intact fungal cells, is mediated by the carbohydrate recognition domain. The Dectin-1 receptor is expressed predominantly by cells of the myeloid heritage (monocytes, macrophages, dendritic cells, Langerhans cells), with the highest levels of expression on inflammatory cells, and cells at the portals of pathogen entry, such as alveolar macrophages (for review see Tsoni and Brown, 2008).

Dectin-1 recognizes particulate and soluble glucans from fungi, bacteria and plants. The minimum unit recognized by this receptor is thought to be a 9- or 10-mer, depending on the method of analysis. In addition, this receptor also recognizes other ligands, including mycobacteria (which do not have glucan) and apoptotic cells.

Lactosylceramide is a glycoprotein containing a hydrophobic ceramide lipid and hydrophilic sacharidic moieties. It recognizes both microbial cells and fungal glucans. Its role is not completely elucidated and will require further study. Some studies suggest the role of an adhesion receptor for pathogens. Lactosylceramide is present in the membranes of many cell types and is particularly abundant in epithelial cells and neutrophils. It has been shown to bind specifically to a variety of microbes and proposed to function as an adhesion receptor between these pathogens and host cells. The interaction of glucan with lacrosylceramide results in cell signaling changes and the subsequent triggering of innate immune responses.

The Scavenger Receptor family is a large heterogeneous group of structurally diverse proteins involved in uptake of modified low density lipoproteins, selected polyanionic ligands and a variety of microbes. Scavenger receptors contain a structurally heterogeneous group of proteins with two transmembrane domains—two intracellular domains and one extracellular domain. These receptors recognize a range of foreign cells, lipoproteins, and polyanions. They can bind lentinan but no specific scavenger receptor for glucan itself has yet been identified. Scavenger receptors were found in a variety of cell types.

Recent studies suggest that these individual receptors that are able to bind glucan collaborate with each other. However, the precise mechanisms still remain unclear. Greater understanding of these interactions, particularly *in vivo*, will be required if the further development of glucans, or the generation of novel therapeutics based on glucans and/or their receptors, is to become a reality.

The activation consists of several interconnected processes that include increased chemokinesis, chemotaxis, migration of macrophages to particles to be phagocytosed, degranulation leading to increased expression of adhesive molecules on the macrophage surface, adhesion to the endothelium, and migration of macrophages into tissues. In addition, glucan binding also triggers intracellular processes that are characterized by the respiratory burst after phagocytosis of invading cells, such as the formation of reactive oxygen species and free radicals. Glucan binding also begins signaling processes that lead to activation of other phagocytes and secretion of cytokines and other substances initiating inflammation reactions (e.g., interleukins IL-1, IL-9, TNF-α). Basically, the binding of glucan to these receptors tricks the neutrophils and NK cells into thinking that the cancer cells are invading yeast cells (full of glucan) and guides them in their destruction.

ACTION

The most pronounced effect of glucans consists of augmentation of phagocytosis and proliferative activities of professional phagocytes—granulocytes, monocytes, macrophages, and dendritic cells. In this regard, macrophages, considered to be the basic effector cells in host defense against bacteria, viruses, multicellular parasites, tumor cells and erroneous clones of our own somatic cells, play the most important role. The effects of glucan on macrophages are schematized in Figure 6. Upon the binding of glucan on the membrane of macrophages, the cells respond by secretion of various cytokines (IL-1, IL-8, and interferon) by proliferation or by production of bacteria-killing molecules.

Macrophages are constituents of the non-specific (innate, non-adaptive) evolutionary older immune system that, beyond phagocytes, is comprised of a complicated family of serum proteins called "complement" and a number of other soluble recognizing and effector molecules.

For the pharmacological effect of glucan, it is important that activated macrophages work not only against the activator but also against any microorganism or tumor cells that are present. Due to the fact that mammals lack β-glucanases, macrophages represent what is probably the only tool for liquidation of glucan in the body. Within the macrophages, phagocytosed glucan is degraded by an oxidative pathway into small soluble fragments that are slowly released as macrophages circulate throughout our bodies settling in numerous organs.

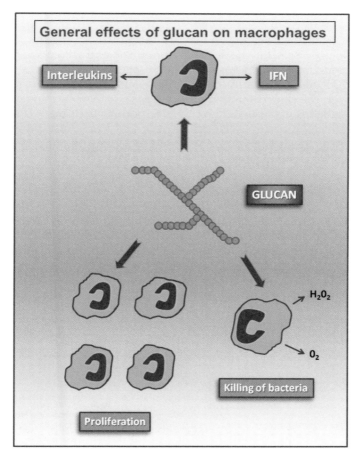

Figure 6

Due to the importance of phagocytic cells (macrophages, neutrophils), any test of potential biological and immunological activities of glucan should start with evaluating the effects on phagocytosis. This decision is based on the fact that phagocytosis is one of the most widely occurring cellular functions. At the same time, phagocytic processes are phylogenetically very ancient. These processes originally served in the normal feeding process of unicellular organisms. However, during the evolution of higher organisms, phagocytosis acquired an entirely new, highly significant *raison d'etre*. Although it is most probable that all eukaryotic cells demonstrate this primitive function; it is especially important for macrophages and leukocytes (such as neutrophils in peripheral blood). Phagocytosis is understood to be one of the most important processes not only in cell biology, but also in immunology, since phagocytosis belongs to the first line of defense. The importance of these processes for the survival of animals and whole species can be demonstrated by findings that not a single creature in the entire animal kingdom was found unable to defend itself by phagocytosis. Moreover, the changes (above all a decrease) in the phagocytic rate of the so-called professional phagocytes are known to be associated with a wide range of diseases. Therefore, it is easy to understand why the question as to whether glucan will have any effect on phagocytosis or not is extremely important.

These miraculous effects of glucan do not end with the activation of immunocytes. In addition to the ability to stimulate the cells of the immune system to perform optimally and maximally, glucan also "cares" about their numbers. It is well established that all cells involved in immune reactions originate from common precursors called stem cells that originate from the bone marrow (see Figure 7). The influx of new cells from the bone marrow is continuous throughout one's entire life. However, the formation and migration of newly

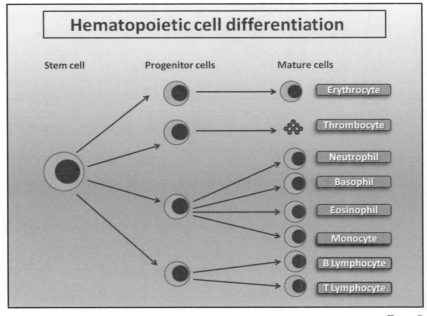

Figure 7

formed cells is limited; and, here again, glucan comes to the rescue. It strongly stimulates the production of precursor cells in the bone marrow, resulting in a more rapid flow of new immunocytes into the bloodstream, and subsequently into the various organs throughout the body. These effects are important not only under normal conditions, as the increased amount of immunocytes in circulation means increased surveillance against potential invaders, but particularly in the case of extreme stress, such as in cancer. With cancer, the limited influx is further reduced by the exhaustion of the immune system and by treatments such as irradiation and chemotherapy.

The immune system consists of various parts, such as, individual cell types and numerous secreted molecules (cytokines, antibodies, complement fragments etc.). It works surprisingly well and is principally responsible for our survival. One has to understand that we are not living in a healthy bubble, but in an environment full of dangerous particles (bacteria, viruses, fungi and parasites) that are constantly invading our bodies. Furthermore, environmental influences such as UV irradiation, antibiotics (both prescribed and in our food chain), electromagnetic fields, or the constantly worsening environment, are causing more or less severe immunosuppression. When we add additional effects such as stress, heavy physical exercise, or simply aging, it is surprising that humans are still reasonably healthy.

As mentioned previously, to be able to fully respond to all invading microorganisms, macrophages need to become activated. Only after several steps of the activation cascade occurred, does a whole sequence of metabolic changes occur simultaneously with changes on the membrane of macrophages. In many cases, the attacking bacteria, either directly or indirectly via toxins such as endotoxin, activates macrophages by themselves. Unfortunately, this natural activation quite often is not enough. Sometimes the bacteria do not adequately activate macrophages; sometimes the whole immune system is weakened, and the number of macrophages is exhausted. Clearly, our immune system needs help; and glucan is an ample provider. It is able to assist our immune system in two completely different ways: first, glucan is taking care of the number of cells involved in immune reactions, and second, glucan activates these cells directly. But more about that later.

The pharmacokinetic parameters of glucan are extremely important for calculating therapeutic dose levels. Although larger glucans appear to be more effective because of their resistance to glomerular filtration and saturation of liver clearance, glucans that are too large have a higher potential for producing undesirable side effects. This condition is especially important when low molecular weight glucans are used. Soluble glucans bind directly to their particular receptors CR3 and Dectin-1 receptor. Soluble glucan polysaccharide, binding to the lectin site of neutrophils or NK cell CR3, generates a primed state of the receptor capable of mediating cytotoxicity of iC3b-opsonized target cells. When this happens, we are speaking of a highly specific stimulation of one particular mechanism of the whole cascade of immune reactions.

In contrast to the water-soluble low molecular weight glucans, glucans that are either insoluble or have large molecular weight can activate immune reactions nonspecifically. Some large glucans trigger a respiratory burst and stimulate the production of various immunoactive substances such as IL-6, IFN and TNF. However, studies performed in recent years showed that insoluble glucan is also internalized by cells that are slowly digested and subsequently released in the form of either soluble glucan or extremely small fragments. The mechanisms of action, therefore, are probably the same as in the case of soluble glucan (Hong et al., 2004).

As mentioned above, glucans are usually considered stimulators or modulators of the cellular branch of immune reaction; and very little attention has been focused on their potential effects on antibody response. We decided to take advantage of the recently published method of evaluating the use of glucan as adjuvant. In our experiments, we injected mice with antigen both with glucan and without glucan. Two weeks later, we measured the level of specific antibodies in the blood of experimental animals. Rather surprisingly, our results showed that most of the tested glucans revealed the stimulation of antibody response—the strongest being Glucagel T and Glucan #300 (Figure 8). Our data suggested that glucan stimulates both branches of the immune system and can be successfully used in vaccines (Vetvicka and Vetvickova, 2007 a).

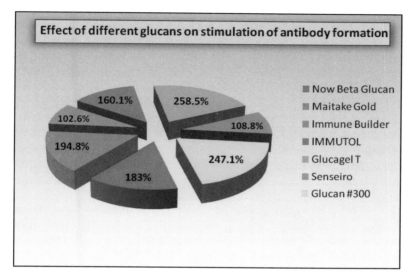

Figure 8

WHAT IS THE ROUTE OF ORALLY-ADMINISTERED GLUCAN?

Our knowledge of the route glucan takes to get into all organs of our body is still limited. The general hypothesis is that so-called M cells in the gastrointestinal tract, and macrophages in the Payer's patches of the gut, bind and subsequently internalize soluble or particulate glucan. Once the glucan is inside the cells, the slow process of digestion occurs, resulting in the production of smaller and smaller glucan fragments. These fragments are later released and subsequently bound by other macrophages. As these cells travel throughout the body, they bring about the glucan fragments which they release in different organs.

Some detailed studies have revealed that after administering oral glucan (either soluble barley glucan or particulate yeast glucan) for three days, labeled glucan within splenic macrophages appeared to be the same size as the starting material. With time, glucan aggregates were found to be concentrated at the edges of the cytoplasm near the cell membrane. Subsequent studies using cultured macrophage cell line examined the fate of labeled glucan added to the cells. These experiments showed that phagocytosed glucan was slowly degraded within cells; and that soluble, biologically highly-active fragments of glucan were released into the surrounding cells. Complete macrophage degradation of glucan required approximately thirteen days. Usually, glucan particles remained intact for four days and appeared to fragment into smaller parts and soluble material during ensuing days (Hong et al., 2004). These results however, do not suggest that we should take glucan only once every thirteen days. Some of the cells of macrophage/neutrophil lineage are rather short-lived, thereby forcing the bone marrow to constantly churn up new cells. In addition, even when it takes as much as two weeks to degrade all ingested glucan, macrophages, in the meantime, will phagocytose additional glucan particles thus guaranteeing a steady release of glucan into the bloodstream.

Still, the number of studies of the gastrointestinal absorption of glucans is limited. Rice's group showed significant differences among various glucans in plasma concentration and in the binding of glucan by gastrointestinal epithelial and GALT cells (Rice et al., 2005). These studies were performed in adult rats with fully developed digestive and immune systems. We tried a different approach. Our focus was on the absorption of glucan during the suckling period when the intestinal barrier function and transport function are not fully established. The immature intestine undergoes dramatic developmental changes of the surface area influenced by a rapid growth of intestinal mucosa. All these changes often lead to the increased susceptibility of developing organism and compromised immune response. Thus, the clarification of the absorption of orally administered glucan by the developing intestine is highly relevant, especially for the potential use of glucan in neonatal and pediatric patients. Results from our studies suggest that only a limited amount of glucan is absorbed by the gut and transferred into systemic blood. In suckling rats, the majority of radioactivity was detected in the gastrointestinal tract and the liver. Thus, we propose that the gastrointestinal epithelium, the GALT cells, and the Kuppfer cells are likely to be the systems that are most affected by orally administered glucan. The enteral administration of glucan during the suckling period could very well be an effective approach in treating gastrointestinal diseases and disorders (Vetvicka et al., 2007 a).

The effects of orally-administered glucans on immune reactions are summarized in the following Table:

Source	Indication	Species	Results	Reference
Yeast	Cancer	Human	Inhibition	Ueno 2000
	Cancer	Mouse	Inhibition	Hong et al., 2004
	Cancer	Mouse	Inhibition	Vetvicka et al., 2002
	Cholesterol	Mouse	Inhibition	Vetvicka et al., 2007 b
Lentinan	Cancer	Mice	Reduction	Kurashige et al., 1997
Schizo-phyllan	Antiviral	Mouse	Increased antibodies	Hotta et al., 1993
SSG	Immunity	Mouse	Increase	Sakurai et al., 1992
	Cancer	Mouse	Inhibition	Suzuki et al., 1991
Maitaki	Cholesterol	Rat	Decrease of Lipids	Kubo and Nanba 1996
	Cancer	Human	Reduction	Nanba 1995
PSK	Cancer	Mouse	Inhibition	Nio et al., 1988
	Cancer	Human	Increased Survival	Kaibara et al., 1976
Agricus blazei	Cancer	Mouse	Enhanced clearance	Ebina and Fujimiya 1998
Sparassis crispa	Cancer	Mouse	Inhibition	Ohno et al., 2000
Seaweed	Cancer	Mouse	Inhibition	Vetvicka et al., 2007 a

7

EFFECTS OF GLUCANS ON CANCER

Cancer is the general name for a group of more than 100 diseases where in abnormal cells in a part of the body begin to grow out of control. Although there are many kinds of cancer, they all start because of the very same reason. Cancer is a leading cause of death worldwide. Approximately 20% of all deaths in the United States are caused by cancer. Generally speaking, cancer is a group of diseases characterized by an uncontrolled growth and spreading of abnormally changed cells. All cancers involve the malfunction of genes that control cell growth, division, and differentiation. Cancer can be caused by many factors, both external and internal. Among external factors are radiation, smoking, various chemicals, and infection factors. Among internal factors are mutations, hormones, and conditions of the immune system. Some of the cancers can be prevented (smoking and heavy use of alcohol can serve as an example). Cancer cell growth is different from normal cell growth. Instead of dying, cancer cells continue to grow and form new, abnormal cells. Cancer cells can also invade other tissues, something that normal cells cannot do. Only about five percent of all cancers are strongly hereditary. The American Cancer Society warns that half of all men and one-third of all women in the US will develop cancer during their lifetimes. The same society suggested that approximately fifty percent of cancer death is caused by cancers that can be prevented.

Despite decades of intensive research, cancer is still one of the most dangerous diseases. In the United States alone, hundreds of thousands of people die every year from various types of cancer. Despite great achievements and decades of intensive, labor-consuming and expensive research, the incidence of various tumors and cancers is still increasing at an alarming rate. Based on the National Cancer

Institute estimates, slightly less than one-in-two men and little more than one-in-three women in the U.S. are likely to contract cancer in their lifetime. The reality is that anyone can develop cancer. The risk of being diagnosed with cancer significantly increases with age. About 77 percent of all cancers are diagnosed in persons aged 55 and over.

The same society estimates that approximately 10.8 million Americans with a history of cancer were alive in January 2004. Some of these patients are completely cancer-free, while others still have some evidence of cancer and have been undergoing a variety of treatments. The American Cancer Society expects about 1,427,180 new cancer cases to be diagnosed in the next year. This enormous number does not even include non-invasive cancers and some types of basal and squamous cell skin cancers (the number of these cancers is expected to reach approximately another 1,000,000/year). Despite all the efforts, about 565,650 Americans are expected to die of cancer every year. This is more than 1,500 people a day. Considering the fact that a full Boeing 747 Jumbo has about 400 passengers, the number of cancer deaths corresponds to the fatal crashes of four Jumbos each and every day.

It has become clear that most cancers have external causes and, in principle, should be preventable. Studies of the association between diet and cancer have demonstrated that the differences in the rates at which various cancers occur in different human populations are often correlated with differences in diet. There is accumulating evidence to suggest that the composition of the diet has great impact on our immune system. Therefore, changing dietary compositions as a tool to improve the immune function is a current research focus. Glucans, present in various foods such as cereals and mushrooms, have been widely used as immunostimulating agents to promote immune responses.

A very important part of treatment belongs to prevention. Regular and thorough screenings by a health care professional result in the early detection and/or removal of precancerous growth. It is extremely important in every cancer case to find the cancer at the earliest stage when the disease is most treatable. In addition to the most common types of treatments such as, chemotherapy, irradiation, and surgery, alternative types of treatments are getting more and more attention from professional physicians and from the scientific community. There is also a strong current of interest for natural therapies as alternatives for harsh conventional treatments amongst members of the general public. Considering all the progress, the 5-year relative survival rate for all cancers diagnosed between 1996 and 2003 is 66 percent, which is up from the 1970s. However, it is important to note that survival statistics vary greatly by cancer type, its stage at the time of diagnosis, sex, as well as the age and ethnical background of the patient. In addition, these numbers represent the percentage of cancer patients who are alive after five years relative to persons without cancer. These numbers, therefore, do not distinguish between patients who have been cured, and those who have either relapsed, or who are still in treatment.

Beyond the devastating impact of cancer on sufferers and their families, lies the huge expense of its treatment and peripheral costs. The National Institute of Health estimates the overall costs of cancer in 2007 at 219.2 billion dollars. Almost 90 billion are used for the actual treatment, 18 billion represent the costs of lost productivity due to the illness, and the remainder (112 billion) represents the cost of lost productivity due to premature death. It is clear, therefore, that this particular disease is extremely taxing on society.

Based on the multiple biological effects of glucan, it is not surprising that this immunomodulator is also involved in the fight against cancer. Despite the fact that most tumors are recognized by the immune system, the antibody response is usually not strong enough to kill a cancer growth. Even a completely healthy immune system cannot adequately deal with fast-growing cancer cells. Glucan is extremely important, as it is able to cooperate with antibodies. After the tumor cells have been recognized as foreign, specific antibodies are released and subsequently bind to the cancer cells. Following this binding of antibodies, the C3 fragment of complement coats the surface of cancer cells. The glucan-primed cells, such as macrophages and specifically NK cells and neutrophils, then recognize these antibody-C3 coated cells and kill them. Without glucan, the destruction would not take place and the situation would be compounded very quickly.

Since the first direct scientific study forty years ago, the anti-tumor activity of glucan has been clearly demonstrated (Nakao et al., 1983). Since these pioneering studies, numerous animal and human trials have shown remarkable anti-tumor activity against a wide variety of different tumors including breast, lung, and gastrointestinal cancer. Since the 1980s, two types of glucan have been successfully used as traditional medicine for cancer therapy in Japan and China. In Japan, glucan is already licensed as a drug effective in cancer treatment. In addition, at least 26 clinical trials are currently under way in the United States as well as in several European countries such as Turkey and France.

Cancer cells are furtive and therefore difficult to kill. These cells have changed their normal characteristics, resulting in their main danger of out-of-control division. They assault our body and constantly try to find a way to overcome the immune system which, after some time, becomes exhausted. Another form of assault uses changed properties of cancer cells that allow them to escape recognition from the immune system cells, resulting in the subsequent destruction of our natural defense mechanisms.

Under normal physiological conditions, numerous types of cells (including macrophages, white blood cells such as neutrophils, dendritic cells and NK cells) are constantly observing these developments. However, their resources are limited and, in the case of cancer, soon become overwhelmed and overpowered. In a healthy body, the defense system successfully fights the constant stream of invading bacteria and other pathogens. However, the healing ability of our body is not endless. Cancer patients have many additional problems such as stress, which have negative effects on the strength of their natural defensive reactions. When the immune system is compromised in any way, there is a greater risk of tumor development. Clearly, our immunocytes need all the help they can get; and for this reason immunomodulators such as glucan are most important.

There are multiple positive effects of glucan in tumor therapy. One is the direct stimulation of macrophages and natural killer (NK) cells. Macrophages form the first line of defense and protect our body against any type of invading cells—including cancer cells. NK cells represent a special subtype of "bloodthirsty" lymphocytes and have an extremely important function—to specifically recognize and kill tumor cells. Together, these cells form a defensive line that guards the integrity of our body. Their job is not easy and considering the fact that they perform this function literally 24/7, it's easy to see how they can become exhausted. Therefore, they can use all the help they can get.

In order to fully investigate the mechanisms and potential

utility of glucan in immunotherapy, it was necessary to develop a suitable mouse system. First, mouse leukocyte CR3 was shown to function as a receptor for glucans in the same way as human CR3. Next, it was shown that the primed state of macrophages and NK cells remained detectable for up to 24 hours after a short interaction with glucan (Vetvicka et al., 1997).

One of the keys to successful immunotherapy of cancer is thought to be the generation of tumor-specific antibodies and tumor-specific cytotoxic lymphocytes. Despite being less aggressive than cells, the occurrence of natural antibodies in tumors has been known for decades. Immunohistochemical staining of excised tumors for immunoglobulins and complement fragments, as well as circulating tumor-reactive antibodies, have been noted with mammary carcinoma (as well as lung and colonic cancers). Antibodies alone will probably not prevent tumor growth. However, they may reduce metastases.

Therefore, we investigated the occurrence of antibodies and complement fragments (most of all C3 fragment) in both the animal and human models. Our investigation showed that the majority of malignant cells in mammary carcinomas are naturally targeted with C3 for cytotoxicity by NK cells bearing CR3 receptors that have been primed with glucan. Both freshly excised human mammary tumors and established breast cancer cell lines were examined and published reports of both circulating antibodies to tumors and tumor opsonization with immunoglobulins and C3 were confirmed. Further, whereas older investigations have tested tissue sections by immunohistochemistry, our investigation examined single cell suspensions of tumors by flow cytometry. This allowed full quantification of antibodies and C3 fragment of complement. Our results suggested that while the majority of malignant cells within tumors bore IgM, IgG and C3, the surrounding normal breast epithelium was devoid of these immune reactants. Similarly, it was reported that tumor section from all 48 tested patients with mammary carcinoma were positive for IgG as well as C3 (Niculescu et al., 1992). Experiments that followed showed that daily therapy, with soluble or insoluble glucan for two weeks, resulted in a 70 to 95 % reduction in tumor weight as compared to the control group. Figure 9 shows the results of experiments comparing the effects of three types of commercially available glucans (yeast-, seaweed- and mushroom-derived glucans) on growth of either lung or breast cancer. From these data, one can clearly see that both yeast-derived and seaweed-derived glucans were significantly more active in tumor growth reduction than glucan isolated from mushrooms. These data correspond with long-term effects of orally-administered glucans on secretion of IL-2 by mouse spleen cells (Figure 10).

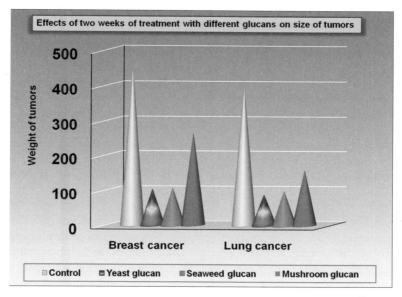

Figure 9

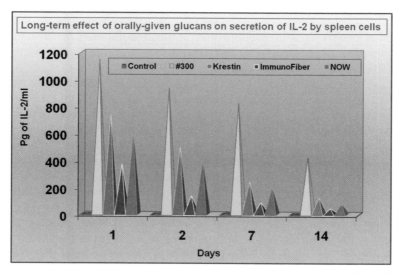

Figure 10

Monoclonal antibodies are currently being evaluated in an increasing number of disorders, including cancer. Although many patients respond to the antibody treatment, remissions are often transient. For example, more than 50% of lymphomas recurrent after rituximab treatment failed to respond the second time. The reasons for this resistance to the antibody treatment are currently unknown but might include loss of antigen, pharmacokinetic variations among individual patients, or resistance to complement activity. It is clear, however, that for a truly reliable antibody treatment, there is a strong need for a synergetic support.

Numerous recent studies have shown that glucan is extremely active in cooperation with antibodies that naturally occur in case of cancer (Hong et al., 2003, 2004). We have to keep in mind that antibodies alone cannot make tumor cells disappear. However, following the binding of antibodies on the surface of cancer cells, C3 fragments of complement coat the cancer cells. The glucan-primed cells, such as blood neutrophils, macrophages, and natural killer cells, then specifically recognize these complement-antibody complexes and kill the tumor cells. Despite the fact that most tumors are recognized by the immune system, the antibody response is usually light and often is not strong enough to kill a cancer growth. And here comes glucan to the rescue: glucan-activated immunocytes recognize and kill cancer cells coated with antibodies. Without the glucan-caused activation of immunocytes, the cancer cells remain coated with antibodies but no killing occurs. It is important to note that similar effects can be also observed with neutrophils.

Neutrophils are usually not active in fighting cancer cells. Their duties are elsewhere: they comprise 40 to 75 % of total while blood count; and they are involved in acute inflammation and active phagocytosis. In addition to the direct killing of invading bacteria, they also participate in such reactions as the uptake of antigen-antibody complexes. However, when these cells and their CR3 are primed with glucan, they can easily kill cancer cells (Hong et al., 2004). Particulate yeast-derived glucan was used in preclinical animal models in combination with exogenous administration of anti-tumor antibodies to test the efficacy for tumor therapy. For example, the combination of daily oral glucan and weekly administration of anti-tumor antibodies caused significant tumor regression of 80% or more compared to treatment with antibodies alone (Hong et al., 2004). Immunotherapy with glucan substantially enhances the therapeutic efficacy of anti-tumor antibodies in the experimental murine breast, lung and lymphoma tumor models. To facilitate translation from preclinical models to clinical application, human carcinoma-challenged xenograft models were further used to test the therapeutic efficacy of combined anti-tumor antibodies with glucan therapy. Another research group confirmed this concept. The group of Dr. Cheung of the Memorial Sloan-Kettering Cancer Center studied a combination of the complement-activating antibody rituximab with glucan against subcutaneous non-Hodgkin's lymphoma. The results of this study were a significant suppression of this lymphoma with a combination of intravenous rituximab and oral glucan as compared to animals treated with antibodies or glucan alone. The survival of mice with disseminated lymphoma was significantly increased in the combination group when compared to other treatment groups. In addition, no clinical toxicity was observed (Modak et al., 2005).

Compared to the traditional treatments of cancer, this type of treatment has one big advantage: it acts without any negative side effect. The experiments mentioned above provided the basis for clinical trials in which patients are treated with antitumor antibodies and oral glucan. Dr. Cheung's research data, in particular, resulted in a series of clinical trials.

Several humanized antitumor monoclonal antibodies (Herceptin™, Rituximab™, Avastin™, Zevalin™, Campath-1H™ and Erbitux™) are now being used to treat patients with metastatic breast carcinoma, non-Hodgkin's lymphoma, chronic lymphocytic leukemia, and metastatic colon carcinoma. Avastin™ (bevacizumab) is particularly successful and has been approved to treat a number of cancers. This antibody targets a growth signal called vascular endothelial growth factor (VEGF) that cancer cells send out to attract new blood vessels. Avastin™ intercepts a tumor's VEGF signals and stop it from reaching its targets.

Because the preliminary experiments on animals were so promising, it is no wonder that these experiments are currently repeated in several clinical trials (among others, by researchers in The Memorial Sloan-Kettering Cancer Center and in the Brown Cancer Center in Louisville). Additional clinical trials focusing on soluble glucan Imprime PGG™ in combination with monoclonal antibody from ImClone System, Erbitux™ (Cetuximab™), and Irinotecan™ (chemotherapy drug from Pfizer) are currently under way. As the preliminary results are already exceeding all expectations, the manufacturer of Imprime PGG™, Biothera, has already started additional cancer clinical trials. Based on the latest information, both current Phase II clinical lung cancer trials have achieved their stage one endpoints, demonstrating a significant level of efficacy among the initial group of patients. Both trials are currently enrolling additional patients at 26 medical centers in the U.S. and Germany. These studies, which will enroll up to 90 patients each, further support the synergy of glucan and monoclonal antibody therapies. In addition, Biothera launched a Phase III clinical trial evaluating the combination therapy of Imprime PGG™ and Erbitux™ in colorectal cancer patients. This study will enroll up to 795 patients and will be conducted at 50 locations worldwide, including the U.S., France and Germany initially.

Additional humanized antitumor monoclonal antibodies Rituximab™ and Campath-1H are other promising targets. Clearly, glucan functions as a potent adjuvant for monoclonal antibody therapy of cancer to elicit novel neutrophil- and macrophage-mediated tumor-killing mechanisms that are not activated by monoclonal antibody therapy alone. Even when glucan by itself was repeatedly shown to cause tumor regression, it was most probably a result of cooperation with natural antitumor antibodies. However, combined therapy with monoclonal antibodies, or vaccines in combination with glucan, offers several advantages: it is easier to elicit an antibody response than with cytotoxic lymphocytes; and the therapy can use humanized antibodies. Therefore they are not relying on the patient's own immune response that is frequently suppressed because of tumor burden and previous irradiation or chemotherapy. The preliminary results are often so promising that glucan-producing companies are acquiring anti-cancer monoclonal antibodies from their developers (a recent Biothera's purchase of a MUC1 anti-cancer monoclonal antibody from Antisoma can serve as an example). More details on current use of glucan in clinical trials are given in following table (Yan, 2011).

GLUCAN CLINICAL TRIALS IN CANCER THERAPY

Tumor Type	Intervention	Phase	Instituted
Neuroblastoma	β-glucan plus anti-GD2 mAb	Phase I	Memorial Sloan-Kettering Cancer Center
Neuroblastoma	Oral β-glucan plus a vaccine containing two antigens covalently linked to KLH	Phase I	Memorial Sloan-Kettering Cancer Center
Chronic Lymphocytic Leukemia(CLL)/ Small Lymphocytic Lymphoma (SLL)	Rutiximab plus oral glucan supplement	Phase II	James Graham Brown Cancer Center
Advanced non-small cell lung carcinoma	β-Glucan MM10-001	Phase I	Beckman Research Inst.
Lymphoma or Leukemia	Rutiximab plus oral β-glucan	Phase I	Memorial Sloan-Kettering Cancer Center
Breast cancer	Soluble β-glucan plus mAb plus chemotherapy	Phase I/II	Biotec Pharmacon ASA
Non-Hodgkin's Lymphoma	Soluble β-glucan plus Rutuximab plus COP/CHOP	Phase I	Biotec Pharmacon ASA
Non-small cell lung carcinoma	Oral β-glucan supplement	N/A	James Graham Brown Cancer Center
Non-small cell lung carcinoma	Imprime PGG in combination with Avastin® (bevacizumab) and two chemotherapeutic agents, carboplatin and paclitaxel.	Phase II	Biothera
Non-small cell lung carcinoma	Imprime PGG in combination with Erbitux® (cetuximab), carboplatin and paclitaxel	Phase II	Biothera
Metastatic colorectal cancer	Imprime PGG™ and Erbitux®	Phase Ib/IIa	Biothera
KRAS-Mutated Colorectal Cancer Patients	Imprime PGG™ and Erbitux®	Phase II	Biothera

The major side-effects of both traditional chemotherapy and/or irradiation are leucopenia (strong decrease in the number of immunocytes) and significant suppression of the immune system. Leukopenia, caused by either chemotherapy or irradiation, is a significant problem in most cancer patients. Both of these negative effects limit the dosage and frequency of treatment. The effects of injected glucan on enhanced recovery after experimental leucopenia have been documented (Turnbull et al., 1999). In the cyclophosphamide-induced leucopenia model, there frequently was a peak at about day three followed by a slow recovery. Similar data were found after irradiation. Numerous studies showed that glucan exhibits a preferred type of recovery without the short-term overstimulation of myelopoiesis (Vetvicka et al., 2007 b). Figure 11 shows glucan-induced recovery of bone marrow cells after sub-lethal irradiation of mice. The data demonstrate that regardless of whether the glucan was used simultaneously with irradiation, or one or two weeks earlier, the bone marrow recovery was significantly improved. Fifteen days after irradiation, the control group (without glucan) still showed lower cell numbers than before irradiation. On the other hand, in all glucan-treated cases, the number of immunocytes in damaged bone marrow returned to the normal levels much faster. These research data might now be more important than ever – as the current situation in Japan

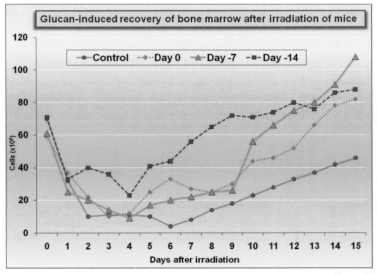

Figure 11

48

has raised the possibility of a radioactive cloud. With β 1,3 glucan we can prepare ourselves for the possible negative effects of radiation, both from the air or from contaminated products.

Cancer patients observed additional benefits of glucan. A recently published study by the research teams from Germany and Turkey found that glucan induces proliferation and activation of monocytes in peripheral blood of patients with advanced breast cancer (Demir et al., 2007). Similar findings were recently observed using yeast-derived glucan in prostate cancer patients (Magnani et al., 2010). These findings indicate that glucan not only helps bone marrow to overcome the negative effects of cancer, chemotherapy, and irradiation, but also increases the biological activities of these newly-formed immunocytes. An interesting approach used glucan to enhance cytostatic activity of a common cytostatic drugs cyclophosphamide, doxorubicin and actinomycin D. In all cases, glucan increase the effects of these chemotherapeutics (Badulescu et al., 2009).

So, is it a good idea to take glucan to prevent cancer? Yes. Glucan will help protect you against cancer and many other illnesses that would normally be eradicated by a strong and fully functioning immune system. But remember, I am a scientist and not in the business of selling glucan. I cannot give any guarantee that you will avoid cancer or any other illness, because there is still a lot we do not know. Even though we can say that glucans are the most significant anti-cancerous immunostimulants that we are aware of, they have not been tested against all varieties of cancers.

Also, we are exposed to carcinogens in our everyday life that are extremely powerful. There are so many of these carcinogens that we cannot take them out completely. Fortunately, our body is able to eliminate a significant part of the constant attack, otherwise we would never reach adulthood. Again, glucan can help our immune system to perform better in the elimination of carcinogenic stress. However, unless we run tests on glucans against all of them to see how effective they are, we will not be able to say that it is a guarantee against everything, although, it is certainly effective against many carcinogens so your risks of cancer would be greatly reduced. It would certainly be a good idea for cancer patients to take glucan as they may find that their cancer regresses or stops progressing. Also, since their immune system is weakened, they are likely to acquire all sorts of viruses and bacteria. Having cancer is hard enough, but having cancer and a serious infection is even worse. Any help we can give our immune system is going to be beneficial.

Similarly, it would be a good idea to take glucan to avoid metastasis, because it would actively help reduce the number of cancer cells. Metastasis might still happen, but they will find it harder to settle, or their onset will be delayed. Remember, none of the anti-cancer drugs offers miracles, and the same is certainly true about glucan. Therefore, every single killed cancer cell counts.

Glucans are extremely beneficial to people undergoing chemotherapy. Chemotherapy drugs are drugs that kill cells indiscriminately. The hope is that they will kill fast-dividing (i.e., cancerous) ones more quickly than healthy ones. However, healthy cells are very much under attack. This is why people become so weak. Their immune system pretty much disappears. Taking glucan will protect your immune system cells while allowing the drugs to do their job on the cancerous ones. The production of white blood cells will start again very quickly after your chemotherapy treatment, thus you will recuperate a lot quicker. Studies have also shown increased survival times when patients took glucans at the same time as chemotherapy as compared to those who did not. So, to summarize: when people take glucans at the same time as chemotherapy, they have less side-effects, do not feel quite as ill, and the effects of chemotherapy are greatly improved. In other words, the treatment is more likely to work. The effects of glucan here are the same as in the case of irradiation.

One side effect all cancer patients facing chemotherapy worry about is hair loss. I am often asked if glucan will prevent it. The answer is that nobody knows because this effect has not been studied. Also, animals do not lose their hair with chemotherapy, not a single one. When compared to the amazing increase of bone marrow production and the speed of recovery, the wig is a minor problem. However, I have one piece of anecdotal evidence that may be encouraging: A good friend of mine who took yeast glucan while being treated for cancer with chemotherapy and radiation did not lose his hair. He also lived two years longer than predicted by his doctors and was relatively pain-free.

Currently, most anti-tumor immunotherapies (except anti-tumor antibody therapy) are reserved for advanced patients that have failed to respond to conventional therapy. The challenges of antitumor immunotherapy lie in many aspects, including immune tolerance established by tumors. Although robust populations of immune effector cells can be generated ex vivo, clinical and pathologic complete responses remain rare in cancer patients treated with these modalities. Combined immunotherapy utilizing glucan and anti-tumor monoclonal antibodies is one means of breaking immune tolerance to tumors. The clinical usage of anti-tumor monoclonal antibodies continues to grow and will soon be incorporated into the standard of care for cancers of multiple organs. Increases in the number of these antibodies offer more opportunities to design versatile combinations with glucan therapy. This modality can be also used in synergy with most tumor vaccines, as long as the anti-tumor humoral responses are elicited, and the antibodies can bind to tumors resulting in complement activation.

Despite the demonstration of glucan-primed neutrophils that mediate CR3-dependent inhibition of iC3b-opsonized tumors *ex vivo*, there are some challenges to the success of combined glucan with anti-tumor monoclonal antibody therapy. Some potential means to overcome these obstacles include breaking the physiological barrier to inflammation by utilizing exogenously administered pro-inflammatory cytokines, or inducing the tumor cells, or stroma, to produce inflammatory cytokines. Similarly, strategies that result in the deposition of more iC3b, including anti-tumor cocktail antibodies that bind to multi-targets or amplification of complement activation, would be expected to improve glucan-mediated therapeutic efficacy. These studies should shed light on how to better design robust and effective combination therapy in cancer.

Two new and potentially important possibilities of glucan action in the fight against cancer have been recently described. Prevention of cancer by vaccination is a long-term dream, because clinical use is still hampered by weak induction of antitumor immune responses. To overcome the limited responses observed so far, vaccines are increasingly combined with an immunological adjuvant able to intensify immune response. A recent study showed that oral administration of glucan during prophylactic peptide vaccination diminished the growth of B cell lymphoma (Harnack et al., 2009). Another interesting and potentially clinically relevant approach is in the form of photodynamic therapy. This type of therapy combines a drug and photosensitizer with a specific type of light to kill cancer cells. In addition to producing necrosis and prompting apoptosis in the tumor, phototherapy also triggers the immune system of the host. Experiments on mice clearly showed that a combination of photodynamic therapy with glucan strongly enhanced the tumor response to the therapy, which resulted in pronounced necrosis of treated tumors and suppression of the DNA damage repair system (Akramiene et al., 2010). From these reports it is clear that the role of glucan in the fight against cancer is wider than originally expected.

8

EFFECTS OF GLUCAN ON INFECTIONS

In reality, the effects of glucan on anti-infection immunity are among the oldest studied effects of glucan. The many papers describing the glucan-induced stimulation of immune reactions after viral, bacterial, fungal and parasitic infection are too many to be mentioned in detail. Virtually all the animal models used have shown that there was not a particular type of experimental infection that would be resistant to glucan treatment.

Glucans have been demonstrated to protect against infection with both bacteria and protozoa in several experimental models and were shown to enhance antibiotic efficacy in infections with antibiotic-resistant bacteria. The protective effect of glucans was evidenced in experimental infection with *Leishmania major* and *Leishmania donovani, Candida albicans, Toxoplasma gondii, Streptococcus suis, Plasmodium berghei, Staphylococcus aureus, Escherichia coli, Mesocestoides corti, Trypanosoma cruzi* and *Eimeria vermiformis.*

Some detailed studies have shown significant synergy of glucan with common antibiotics. Original studies done on guinea pigs demonstrated that simultaneous administration of glucan and antibiotics elevated the ability of animals to resist lethal septic infection by antibiotic-resistant bacteria (Kernodle et al., 1998). At the same time, those results suggested that the use of glucan can help to lower the doses of antibiotics in commercial farming, which is particularly important since there is a strong effort to completely abandon the use of antibiotics in all farmed animals. Positive effects were also found in patients after cardiopulmonary bypass (Hamano et al., 1999).

In addition to hundreds of reports on glucans stimulating the immune response against many infections, there have been numerous studies, including clinical trials, conducted with glucan and infections in humans. Alpha-Beta Technologies conducted a series of human trials in the 1990s. Using double blind, placebo-controlled trials, these studies showed that patients who received glucan had significantly less infections, had a decrease in the use of antibiotics, and a shorter stay in the intensive care unit (Babineau et al., 1994).

Evidence from animal studies demonstrates that beta-glucan can reduce the amount of conventional antibiotics required in infectious conditions such as peritonitis (inflammation of the membrane lining of the abdominal and pelvic cavities). In mice infected with bacteria to induce peritonitis, a combination of glucan and standard antibiotics increased the long-term survival by 56%. Bacterial counts were noticeably down within eight hours of the injection, and the numbers of key immune cells were markedly higher.

Although the anti-infective properties of glucan have already been established, the majority of the work has been done with parentally administered glucans and the number of bacteria tested was limited. Due to the recent threats of bioterrorism, it was particularly interesting to see the strong prophylactical effects of orally-delivered glucan on infection with *Bacillus anthracis* (Vetvicka et al., 2002).

Browder et al. (1990) described stimulation of human macrophages in trauma patients and found that glucan therapy strongly decreased septic morbidity. A multicenter, double blind study found the optimal dosage of glucan in high-risk surgical patients. In addition, these studies demonstrated the safety and efficacy of glucan in surgical patients who underwent major thoracic or abdominal surgery. Since no adverse drug experiences associated with glucan infusion have been found, glucan-treated patients had significantly lower levels of infections. The biological effects of glucan on anti-infectious immunity are two-fold: macrophages are activated to produce substances (such as H_2O_2) directly killing the bacteria and stimulation of B lymphocytes to produce more antibodies.

Sepsis leads to the damage and dysfunction of various organs. One of the underlying mechanisms is thought to be the oxidative damage due to the generation of free radicals. Sener's group investigated the putative protective role of yeast-derived glucan against sepsis-induced oxidative organ damage. Sepsis was induced by caecal ligation and puncture in rats. Sham operated (control) and sepsis groups received saline or glucan once daily for 10 days and 30 minutes prior to and six hours after the puncture. Sixteen hours after the surgery, the rats were decapitated and the biochemical changes were determined in the brain, kidney, heart, liver and lung tissue. Tissues were also examined under light microscope to evaluate the degree of sepsis-induced damage. The results demonstrate that sepsis significantly decreased GSH levels and increased the MDA levels and MPO activity causing oxidative damage. Elevated plasma TNF-alpha levels in septic rats significantly reduced to control levels in the glucan-treated rats. Since glucan administration reversed these oxidant responses, it seems likely that glucan protects against sepsis-induced oxidative organ injury (Sener et al., 2005).

Some experiments are nearing clinical application in humans, such as a study of the effects of glucan against oxidative organ injury and sepsis. Recent studies also showed the protective effects on lung injury after experimental caecal punctures in rats. It is particularly important to know that similar data were obtained where glucan is already widely used utilizing commercially important species of animals such as pigs, chickens, horses, and fish. A similar experimental model was used also for evaluation of the possible cooperation of glucan with antibiotics. The study used two different (soluble and particulate) glucans, either alone or in combination with antibiotic therapy, in the evaluation of their ability to augment the survival of rats following caecal ligation and puncture. The study concluded that both glucans acted synergistically with antibiotics in polymicrobial sepsis (Bowers et al., 1989).

It is curious that much less is known about the role of glucan in fighting viral diseases. Initial studies on the antiviral effects of yeast glucan were undertaken by Williams and DiLuzio (1980). Their studies showed pronounced survival of mice lethally challenged with murine hepatitis virus, regardless of whether glucan was used before or after the infection. Further studies showed significantly better protection against equine encephalomyelitis virus and *Herpes simplex* Type II encephalitis in mice. Additional studies showed a decrease of viral infection using lentinan, SPG, and bacterial curdlan, which are different types of glucan. Glucan increased survival of cells infected with hepatitis virus strain MHV-A 59 in mice. There are some data from human studies showing positive effects on HIV-infected patients (Itoh et al., 1990). In addition, during a placebo-controlled trial in

San Francisco, HIV-positive patients experienced an increase in the number of CD4 lymphocytes after glucan treatment. These effects were most pronounced when glucan was used simultaneously with other antiviral agents (Gordon et al., 1995). Another group studied the effects of co-administering viral proteins with glucan on the production of immune responses. Experiments on mice showed that animals obtaining the conjugate of glucan with viral proteins had significantly higher antibody titres (Mohagheghpour et al., 1995). Lentinan, glucan approved for human medicine, has been found to improve the immune effects of Newcastle disease in chickens (Guo et al., 2009). Recently, an interesting study showed that glucan increased the effects of anti-rabies, and anti-parvovirus vaccine in dogs (Haladova et al., 2011). This suggested not only a possible use in small animal clinical practice, but also further supported the idea that glucan is similarly active in the fight against viral infections as it is in the fight against bacterial infections.

9

EFFECTS OF GLUCAN ON BLOOD SUGAR AND CHOLESTEROL

In the United States alone, approximately 500,000 people die of heart disease each year. The link between elevated cholesterol levels and the risk of coronary disease has been clearly established. In addition, cardiovascular disease related to elevated blood cholesterol levels is still the most common cause of death in humans in western countries. Since none of the current cholesterol-lowering drugs are without side-effects, the search for a natural modulator of cholesterol concentrations is an important task. Once again, glucan comes to the rescue.

In addition to the effects of glucan oriented towards the immune system, glucans were also shown to reduce the total and LDL cholesterol levels of hypercholesterolemic animals and patients. Nearly 50 years ago, the possible effects of dietary fiber were first suggested by Keys, and these effects were later found to be associated with glucans (Tietyen et al., 1990). The cholesterol-lowering effects of fibers are routinely associated with β-glucans. Due to the high consumption of oats, or oat bran, most of the attention has been focused on the relationship of oat-derived glucan to cholesterol levels in both animals and humans.

Oat-derived β-glucan significantly improves HDLC and diminishes LDLC and non-HDL cholesterol in overweight individuals with mild hypercholesterolemia (Reyna-Villasmil et al., 2007). There are, however, some studies where no such effects were found. Due to the fact that glucan can be easily added to the food during the preparation of oats, the experiments testing the role of glucan in lowering cholesterol are usually done with oat-derived glucan. At the same time, there is no real reason to expect that yeast-derived glucan will not have the same effects.

Despite extensive investigations using not only various types of glucans, but also numerous concentrations and routes of administration, including oral, intraperitoneal, subcutaneous and intravenous applications, there is no conclusive report explaining the mechanisms of action. Without any direct evidence, these effects of glucan are usually described as the result of fiber intake and subsequent decreased absorption of bile acids. However, most of these studies suffer from the fact that they did not evaluate the effects of isolated glucans. They used only crude extracts without any knowledge as to whether these glucans are even digested.

Therefore, we felt the need to directly compare different types of glucan. We chose four glucans widely sold and available in the US, Europe and the Far East, representing grain, mushroom and yeast-derived glucans in soluble and insoluble form. Briefly, *Glucan #300* is an insoluble yeast-derived glucan; *Krestin* is a soluble mushroom-derived glucan; *ImmunoFiber* represents a soluble grain-derived glucan; and *NOW* is a mixture of both insoluble glucans from yeast and soluble glucans from mushrooms. Our study was not only the first to directly compare the cholesterol-lowering activity of several different glucans but was also the first to compare control animals and mice with experimentally-induced cholesterolemia.

First, we studied the effect of long-term feeding with a diet supplemented with glucan. Our data showed strong time-dependent effects of *Glucan #300* and *Krestin* glucans on lowering cholesterol. The effects of other glucans were less pronounced and in the case of *ImmunoFiber* we found almost no effect (Figure 12).

The mice were then given a diet with added cholesterol. The blood cholesterol levels obtained after two weeks of cholesterol feeding were used as a positive control. The cholesterol-rich diet was followed by 40 days of feeding with a glucan-rich diet. Individual groups of mice were sacrificed in 10-day-intervals and their cholesterol levels were evaluated. The results, summarized in Figure 13, showed that during short time intervals, all glucans lowered the cholesterol levels in hypercholesterolemic animals, but in the long term only

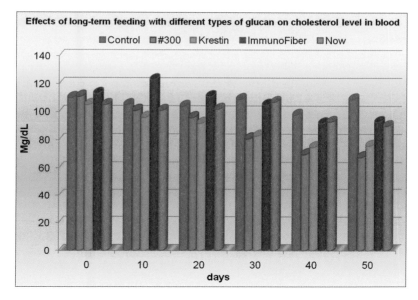

Figure 12

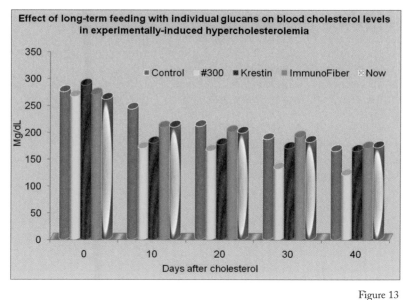

Figure 13

Glucan #300 glucan retained this activity. The usefulness of glucan in lipid lowering was further confirmed by a recent meta-analysis of randomized, controlled trials (AbuMweis et al., 2010).

Similar activity was also found with yeast-derived glucan *Betamune*. The fact that both glucans are insoluble indicates the lack of relevance of the solubility. The fact that both soluble and insoluble glucans are processed by macrophages into small fragments that subsequently prime cells for biological activity (Hong et al., 2004), further stresses the lesser (if any) significant role of glucan solubility. In addition, the effects of both types of glucans suggest that the general mechanism is most probably acting via increased intestinal viscosity, causing the reduction of cholesterol absorption and ending with subsequent cholesterol excretion. Recently published meta-analysis of the effects of glucan intake on blood cholesterol compared thirty research articles and confirmed that consumption of 3 grams per day of glucan is sufficient to decrease blood cholesterol (Tiwari and Cummins, 2011).

The effects of glucan on blood sugar are less studied. Older studies suggested that glucans might reduce blood glucose concentrations after eating, possibly by delaying bowel movements so dietary glucose is absorbed more gradually. Some studies have shown the hypoglycemic activity of natural glucans. Additional studies have demonstrated the strong hypoglycemic activity of synthetic polysaccharides. Some follow-up experiments even suggested that small synthetic oligosaccharides lower the blood sugar similar to natural glucans, which would rule out the possible effects via fiber action. However, the mechanisms remain unknown.

Some groups suggested similar mechanisms to those involved in the lowering of cholesterol, ie., changes in the increase of viscosity of the alimentary bolus and changes in gastric emptying. A very interesting study from Japan evaluated glucan's possible effect in preventing the development of diabetes and insulitis. Using a model of rats with a high occurrence of spontaneous diabetes, the authors showed that the administering of glucan decreased the cumulative incidence of diabetes from 43.3 percent to 6.7 percent and the incidence of insulitis from 82.4 percent to 26.3 percent. Eight out of nine rats were completely free from diabetes for five weeks after glucan treatment was discontinued. The authors concluded that glucan can modulate the autoimmune mechanisms directed to pancreatic islets and inhibit the development of diabetes (Kida et al., 1992).

It is clear that glucan (most often from oats) can be linked to cholesterol reduction and the science is sufficiently robust to have merited health claims in a range of countries. Indeed, such a claim has existed in Sweden since 2002, in the UK since 2004, while the French Food Health and Safety Agency (AFFSA) approved a health claim for glucan cholesterol-lowering in 2008. In May 2008, the US Food and Drug Administration (FDA) added certain oat products to a health claim linking soluble fiber and risk of coronary heart disease.

Therefore, we focused our attention on the effects of glucan administration on the levels of blood sugar. Feeding with glucan did not significantly affect the sugar levels. However; a different situation was found when we used mice with experimentally-induced hyperglycemia. After two weeks of feeding, *Glucan #300* significantly lowered the sugar levels to almost normal (Figure 14). A longer application of glucan resulted in additional significant activity of *ImmunoFibre* glucan. Experiments describing the effects of glucans on cholesterol and blood sugar are described in full detail here (Vetvicka and Vetvickova, 2007 b).

From these experiments, we can conclude that with respect to natural glucans there is a yes-or-no effect suggesting that highly purified and highly active glucans will have a pleiotropic impact. Whereas, poorly isolated and/or less active glucans will have only mediocre biological properties.

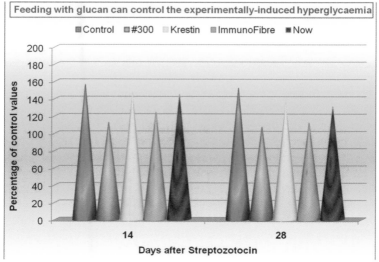

Figure 14

61

10

GLUCAN AS PREBIOTIC

Prebiotics are non-digestible food ingredients that help to stimulate the growth and activity of bacteria in the digestive system. The whole idea was introduced in 1995 by (Gibson and Riborfroid 1995). The second term, sometimes mixed with the term "prebiotics," is probiotics. These are live microorganisms that are beneficial to the organism. The most common probiotics are lactic acid bacteria *(Lactobacillus acidophillus)* and bifidobacteria *(Bifidobacterium lactis)*. These bacteria are usually consumed as part of fermented food with added live cultures. The use of yogurt is the most common example. The nutraceutical potential and bioactive properties of polysaccharides including glucan have been investigated in detail during last decades.

The most studied prebiotics are oligosaccharides, primarily fructooligosaccharides (oligofructose and inulin) and galactooligosaccharides. Despite significant efforts, the single molecule responsible for these effects has not been found, although saccharidic chains of 2-64 (sometimes 9-64) saccharide molecules are considered to be the most active. These short chains seem to be smaller than regular, often insoluble glucan. However, we must remember that our body will, slowly but steadily, transform even the largest glucans into small, soluble fragments. Despite the fact that numerous foods contain prebiotics, we cannot call them prebiotics, as **no food is a prebiotic**. As more and more attention has been focused on prebiotics, it is not surprising that scientists also considered glucan. Not surprisingly, most of these observations were published quite recently.

A study on fish showed that using glucan as a probiotic supporting the activity of *Lactobacillus* significantly lowered mortality from *Aeromonas* challenges (Ngamkala et al., 2010). However, more has been done on the prebiotic front. An interesting study showed that addition of glucan and starch during cold storage strongly increased survival of bifidobacteria strains in yogurt, most probably due to the protective effects on bifidobacteria stress by low temperature (Rosburg et al., 2010).

In calves, feeding with tylosin and glucan as prebiotics has positive effects on selected humoral immunological parameters, including the total protein and gammaglobulin concentration (Szymanska-Czerwinska and Bednarek, 2011).

The prebiotic potential of glucan was also tested on humans. A randomized, double-blind, placebo-controlled clinical study was aimed to evaluate the in vivo prebiotic potential of glucan. Fifty-two healthy volunteers were assigned to consume daily a glucan or placebo for 30 days. In volunteers over 50 years of age, glucan induced a strong bifidogenic effect and increase of bifidobacteria. The authors of this study concluded that the daily intake of a glucan was not only well-tolerated, but demonstrated a strong bifidogenic properties in older healthy volunteers consuming their usual food (Mitsou et al., 2010).

A more detailed study compared a prebiotic activity with a structure of glucan. Nine probiotic strains of bacteria were used, with two structurally different versions of the same glucan. These probiotics showed different growth characteristics based on types of glucan used, showing that these effects are not restricted to only one type of glucan. Readers seeking more details about different glucans as prebiotics should read an excellent review written by Synsytsya et al. (2010). It should be noted that the beneficial effects of glucan on microflora were considered high enough to result in a clinical trial of beta glucan in polypectomized patients which is currently under way in Greece.

11

OTHER BIOLOGICAL EFFECTS

One of the strengths of glucan is that it effects numerous cell types and body processes. In addition to the effects on various immune cells (such as macrophages, neutrophils, NK cells etc.) and the effects on blood sugar and cholesterol level (**Chapter 9: Effects of Glucans on Blood Sugar and Cholesterol**) previously mentioned, some additional effects of glucan have been reported. One example is the use of *Wellmune*™ *WGP* glucan on reducing fatigue from physical and psychological stress. Similar data were later produced using oat glucan.

Moreover, glucan has been shown to act as a strong adjuvant in collagen-induced arthritis. These data help to explain experiments that show that glucan functions as an adjuvant for monoclonal antibody therapy by recruiting tumoricidal granulocytes as killer cells (Hong et al., 2004). Recently, glucan has been found to significantly potentiate the methotrexate treatment of adjuvant arthritis (Rovensky et al., 2011). Our own recent data compared the wound healing ability of four different types of glucan – *Glucan #300*, *Glucagel*, NOW Glucan and Epicor (these glucans will be described later). We used the well-established model of scratch wound of a monolayer of human cell line HaCaT. Our results summarized in Figure 15 showed that all glucans supported the wound healing processes. However, our results again confirmed that not all glucans are created equal, with *Glucan #300* showing the strongest activity.

Glucan was repeatedly shown to increase the wound-healing ability of skin. An interesting study showed the effectiveness of a glucan-collagen mixture for treatment of partial-thickness burns

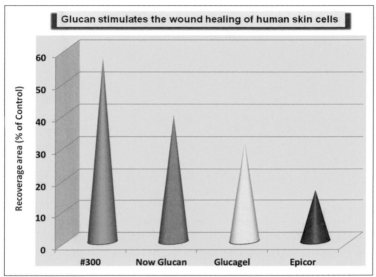

Figure 15

in children. These claims have already resulted in a commercially available glucan-collagen matrix that combines glucan with collagen and has proven to have excellent results in successful treatment of burns in children (Delatte et al., 2001).

Glucan, an immunomodulator, has been reported to increase collagen deposition and tensile strength in experimental models of wound repair. Older data suggested that glucan modulates wound healing via an indirect mechanism in which macrophages are stimulated to release growth factors and cytokines. However, recent data have shown the presence of glucan receptors on normal human dermal fibroblasts which suggests that glucans may be able to directly stimulate fibroblast collagen biosynthesis. Some studies have shown that procollagen mRNA was increased in glucan-treated normal human dermal fibroblasts when compared with the untreated fibroblasts. Collagen synthesis was increased at 24 hours and 48 hours following glucan treatment of normal human dermal fibroblasts.

Although the macrophage is important to wound healing, research has also focused on its relationship to fibroblast and collagen synthesis. A study was designed to assess the effects of enhanced macrophage function on early wound healing, prior to established collagen synthesis. Sprague-Dawley rats had dorsal incisions after three different treatment regimens. In all cases, glucan, given both intravenously or topically, enhanced wound healing. The authors of the study concluded that enhanced macrophage function increased early wound breaking strength. This effect appears unrelated to collagen synthesis but may be related to an increased cross-linking of collagen. In addition to the stimulation of collagen biosynthesis, glucan also improved the wound tensile strength (Portera and Love, 1997). Some additional studies showed stimulated tissue granulation, enhanced re-epithelialization, and an increase in macrophage participation.

The effects of glucan on skin health are most probably the result of interaction of glucan with Langerhans cells within the epidermis. These cells, which belong to the macrophage family, are very sensitive to environmental factors including UV light. As we age, our skin is exposed to an excess of UV radiation, poor nutrition, or environmental toxins, and the protective ability of Langerhans cells to maintain healthy skin is compromised. The glucan-induced protection is a combination of direct activation of Langerhans cells and free radical scavenger capacity. In addition to Langerhans cells, glucan also interacts with skin fibroblasts, further increasing the wound-healing abilities (Kougias et al., 2001). These findings are particularly important for patients with diabetes who are suffering from healing problems. A recent study showed that yeast-derived glucan improved wound healing in diabetic animals (Gulcelik et al., 2010).

A study of mice demonstrated that oral administration of glucan inhibited development of atopic dermatitis (Sugiyama et al., 2010). These experiments prompted the use of glucan in cosmetics. The clinical study of 150 women with topical application of glucan resulted in 27% improvement in skin hydration, 29% improvement in dryness of the skin, and 56% improvement in facial wrinkles. These experiments are summarized in Figure 16.

An additional clinical study of 27 subjects evaluated the effects of glucan on facial fine lines and wrinkles. After 8 hours of treatment with 0.5% glucan solution, 28% of the applied glucan entered the skin and as much as 4% reached the dermis (i.e. the layer where wrinkles form). After 8 weeks of treatment, digital image analysis of silicon replicas indicated a significant reduction of wrinkle depth, height, and overall roughness. The authors concluded that glucan is capable of both entering the epidermis and penetrating deep into the skin thus

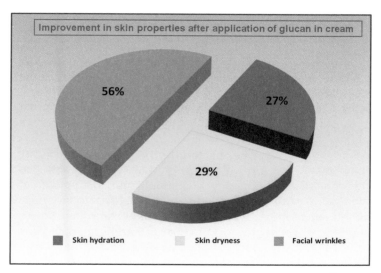

Improvement in skin properties after application of glucan in cream

56%

27%

29%

■ Skin hydration   ■ Skin dryness   ■ Facial wrinkles

Figure 16

delivering significant skin benefits. The observed effects are most probably mediated by fibroblast stimulation and collagen deposition in the dermis (Pillai et al., 2005).

In Norway, Biotec Pharmacon is running clinical trials of their soluble glucan for the treatment of diabetic ulcers. Two parts of the phase III study are currently under way. The same company is also running clinical trials concerning prevention or treatment for patients with oral mucositis which is a painful and potentially serious side effect of radiotherapy. These trials are clearly based on experimental studies showing that glucan reversed experimentally induced damage by irradiation and chemotherapy. In both cases, the results should be known later this year.

Some interesting experiments have shown that glucans of different origins have an effective protective activity against different mutagenic agents such as methyl methanesulfonate and 22-aminoanthracene. Glucans protected the cells by sheltering their DNA and increasing the repair of the double-strand breaks (for review see Mantovani et al., 2007). Additional studies have also proven the efficacy of glucan in reducing the damage caused by other mutagenic agents such as chemotherapeutic drugs cyclophosphamide, adriamycin (better known as doxorubicin), and cisplatin. The protective effect is attributed to the ability of glucan to trap free radicals produced during the biotransformation of these drugs.

All the above effects of beta-glucans can only lead to high expectations and understandable excitement on the part of patients. So if you are taking beta-glucans, you probably wonder how soon it will start working for you. Usually, about 10-to-15 days after you start taking it, your bone marrow will begin to pour white blood cells into your body. These will patrol your entire body to eliminate anything suspicious such as germs, bacteria, viruses, cancer cells, fungus, and parasites. Therefore, if you know that you are going to have radiotherapy or chemotherapy, you should start taking beta glucans at least 15 days prior to your treatment and at least 30 minutes before eating. It is most important that people understand that I am talking about doses of 500 mg to 1 000 mg per day and in the purest form of beta glucan. Many products on the market have no therapeutic value at all because they only contain small amounts of beta glucans per tablet or they are improperly processed.

Thimerosal has been used as a wound disinfectant and a preservative in vaccines for quite some time now. However, studies performed during the past several decades have clearly established the immunosuppressive effects of various types of mercury. Immunodepression was originally connected more with organic methyl mercury. Only later did reports show that inorganic mercury caused similar problems.

Since glucans were repeatedly shown to overcome the immunosuppression caused by either irradiation or chemotherapy, we elected to test the hypothesis that orally-administered glucans can help to overcome mercury-induced immunosuppression. For that particular study, we chose four different glucans that are widely sold and available in the US, Europe, and the Far East. These represent grain, mushroom, and yeast-derived glucans in partly soluble and insoluble form. Briefly, *Glucan #300* is an insoluble yeast-derived glucan; *Glucagel* is a barley-derived glucan; *NOW Glucan* is a mixture of both insoluble glucans from yeast and soluble glucans from mushrooms; and *Epicor* is a partly soluble yeast-derived glucan. In this manner, we covered all the basic types of glucans.

Two weeks of a daily dose of mercury acetate corresponding to approximately 800 µg of mercury/kg (or 200 µg of mercury/kg in case of thimerosal) induced a systemic suppression of all the tested immune reactions—from cellular (phagocytosis, NK cell activity, mitogen-induced proliferation and expression of CD markers) to humoral immunity (antibody formation and secretion of interferon—γ interleukin 6 and interleukin 12).

The effect of individual glucans on the restoration of mercury-induced suppression is given in Figure 17. Exposure to mercury for two weeks severely depressed phagocytic activity. However, when used in conjunction with glucans, it was restored by 71.5% in the case of Hg acetate exposure and by 63.5% in the case of Thimerosal exposure. We began our study by measuring the effects of mercury on the phagocytic activity of the blood neutrophils. Our data showed that only glucans *#300* and *NOW* exhibited a significant stimulation of phagocytosis in control samples. In mercury-induced depression of phagocytosis, only glucan *Glucan #300* was able to help and restore this activity to its normal level.

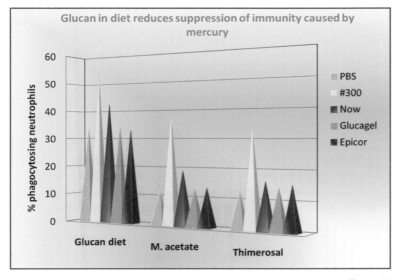

Figure 17

Studies on the simultaneous administration of mercury compounds and glucans began with the changes in mercury's direct toxicity. In both cases, glucans significantly lowered the toxicity. As the direct toxic effects of mercury compounds are hypothesized to be caused by apoptosis, the well-known inhibition of apoptosis caused by glucan might be the explanation. We then continued with groups receiving both mercury and glucans, and we found the same effects for all additional tests, ranging from mitogen-induced proliferation of T and B lymphocytes up to antibody response and NK cell activity. In all cases, we found significant suppression with 14 days feeding of mercury compounds that, for the most part, confirmed previous data obtained by other groups. In all cases, we found that this immunosuppression was partly reversed by glucans. In addition, yeast-derived glucan #300 was consistently the most active glucan.

The answer remains unclear regarding the mechanisms by which glucan reverses mercury-mediated immunosuppression. In addition to the direct stimulation of cells via Dectin-1 and CR3 receptors, glucans are known to alter some important genes and their transcription factors. Moreover, since mercury is known to cause inflammation and oxidative stress, the intracellular mechanisms that involve antioxidant processes might be involved as well. **In conclusion, we report, for the first time, that mercury-caused immunosuppression can be at least partially reversed by the oral administration of glucans.** Based upon our data, one can envision glucan being used as a prophylactic in mercury poisoning.

More and more attention has recently been focused on the role of glucan in inflammation. Some additional, yet lesser known, effects of glucan include the attenuation and even the prevention of experimental colitis. Several experimental studies showed that orally-administered glucan caused significant inhibition of inflammation, production of pro-inflammatory cytokine interleukin 1 beta, and colon shortening. Some studies have shown that glucan plays an important role as an immunomodulator in the treatment of ulcerative colitis (Nosalova et al., 2001). Inhibition of antiviral activity has been found in HIV-infected patients (Itoh et al., 1990). Dietary beta glucan was found to regulate the levels of inflammatory factors and cytokines, reducing symptoms of the inflammatory bowel disease. Similar findings were confirmed on the model of Zucker fatty rats, where glucan-enriched diet was found to prevent the inflammatory state of the metabolic syndrome (Sanchez et al., 2011). In addition, yeast-derived glucan inhibited a dual mechanism of peroxynitrote stroke.

The same type of glucan was also shown to reduce infection-stimulated periapical bone resorption (Stashenko et al., 1995). Positive effects were found in patients after cardiopulmonary bypass (Hamano

70

et al., 1999). In addition, glucan phosphate has been found to induce cardioprotection, most probably by modulating Toll-like receptor mediated signaling (Li et al., 2004).

Scientists in Norway evaluated the effects of glucan on periodontal disease and found that orally administered, yeast-derived glucan significantly reduced periodontal bone loss. Enhanced plasma levels of the HPA axis-driven hormone corticosterone and TGF-1 was also shown in experimental animals. The authors speculate that glucan could possibly be a good candidate for treatment of periodontal disease (Breivik et al., 2005).

So far, our knowledge of possible effects of glucan on allergies is still very limited; and by no means can glucan be considered to be an anti-allergy application. Nevertheless, an interesting announcement came from one of the glucan-producing companies Biothera. In 2001, the company presented results of a placebo-controlled, double-blinded clinical trial of 48 people subjected to high pollen counts. A group consuming yeast-derived glucan demonstrated significant reduction of overall allergy symptoms and severity, reduction in key nasal and eye-related allergy symptoms, and improvements on the Quality of Live Index.

Some interesting findings also came from South Korea. Their scientists studied the effects of glucans isolated from *Aureobasidium pullulans* on inflammation. Different doses of glucan were administered orally to xylene-treated mice; and the changes in the affected areas of their ears were evaluated. The results were compared with the effects of known anti-inflammatory drugs, such as, diclofenac or dexamethasone. They showed that all histological characteristics of acute inflammation, such as severe vasodilatation, edematous changes of the skin, and the infiltration of inflammatory cells, were dose-dependently decreased by glucan treatment, proving just another beneficial effect of glucan (Kim et al., 2007).

A series of very interesting experiments was conducted in Turkey where glucan is manufactured and sold by the pharmaceutical giant Mustafa Nevzat. One study used glucan for the treatment of patients with allergic rhinitis, a disease caused by an IgE-mediated allergic inflammation of the nasal mucosa. The results of the study showed that levels of interleukin 4 and 5, which are responsible for the allergic inflammatory response, were decreased with glucan treatment, while the levels of interleukin 12 were increased. Moreover, the eosinophils, which are important effector cells of the inflammatory response, were decreased. In summary, glucan might play a role as an adjunct to the standard treatment of patients with allergic rhinitis (Kirmaz et al., 2005). However, our knowledge of possible effects of glucan on allergies is still very limited, and by no means can glucan be considered to be an anti-allergy application. These studies were later confirmed with *Glucan #300*.

Glucan was also used in protection against oxidative organ injury. The scientists used a rat model of sepsis using a gut ligation and puncture. The results showed that 10 days of glucan treatment significantly reversed oxidant responses (increase of the MDA levels and MPO activity), demonstrating significant protection against sepsis-induced oxidative organ injuries (Sener et al., 2005). The follow-up studies showing that glucan holds the most promising results to improve the outcome of trauma patients are no longer surprising (Spruit et al., 2010).

Using oat glucan, the researchers first showed that fatiguing exercise has been associated with a decrease in certain functions of neutrophils, which can be offset by the consumption of glucan (Murphy et al., 2007). The same group later tested mortality trends over 30 days in mice run to volitional fatigue on a treadmill for 3 consecutive days and they then inoculated them with herpes simplex virus following 10 days of drinking water with glucan compared to plain water. The results showed a surprisingly high survival of glucan-treated animals (Murphy et al., 2008).

Glucan was recently found to have an additional use: the regulation of stress. We measured the effects of various types of glucan on the levels of stress-induced corticosterone. As experimentally induced stress, we used either restraint or cold. Our results showed that glucans *Glucan #300* and *NOW* successfully helped to keep the stress hormone corticosterone at almost normal levels (Figure 18).

An additional group tested a different type of stress: the oxidative stress of hepatocytes. Evaluating the genotoxic and cytotoxic effects of various substances on freshly isolated hepatocytes, the protective effects of glucans were demonstrated (Horvatova et al., 2008).

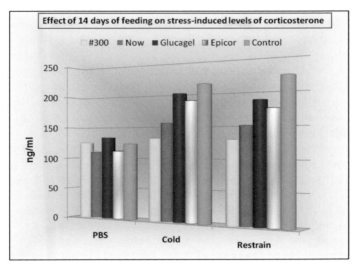

Figure 18

It is also important to note that glucan has significant antioxidant properties. A comparative analysis of lipid peroxidation in phosphatidylcholine liposomes induced by OH radicals showed that a good quality glucan is comparable to two well-established antioxidants—mannitol and tocopherol (Babincova et al., 2002).

A Japanese group from the Japan Women's University focused on the potential role of glucan in iron absorption. Using a rat model employing radioactively labeled Fe, the authors surprisingly found that feeding animals with glucan samples resulted in iron absorption increases in both blood and the small intestine (Mai et al., 2002). Since this is the first paper dealing with this potentially very interesting and clinically important question, the explanation of the possible mechanisms is not clear. The authors speculate that the glucan can bind with iron to form a very weak bond, possibly enabling the transport and release of iron at enterocytes. The higher absorption in the small intestines seems to support this hypothesis. However, these findings need to be repeated before we can safely recommend the use of glucan for people suffering from iron deficiency.

Another potential use of glucan is in drug delivery. Glucans have been used in drug delivery systems either as an actual drug carrier, an adjuvant, or in combination with other materials to form suitable drug delivery systems (i.e. nanogels) and as stabilizer in microcapsules and nanocapsules. For example, pullulan was used in combination with water soluble polymers (such as modified cellulose, starch, and carrageenan) to prepare ingestible films that can contain pharmaceutical, cosmetic, or biologically active agents. Curdlan has been used in the preparation of tablets containing theophylline—a drug used in treatment of respiratory diseases such as asthma. These formulations were prepared from spray-dried particles of curdlan and theophylline. The effect of curdlan on controlled release of the drug after oral administration was then investigated. Scleroglucan has been applied as a coating in liposome formulations and in the preparation of hydrogels, because natural polymers are advantageous over synthetic polymers in the preparation of hydrogels for drug delivery due to their low toxicity and biocompatibility.

Soluble glucans have been studied for the encapsulation of macromolecular drugs such as DNA and proteins. Similarly useful are glucan particles, since glucan derived from yeast can be processed into hollow, highly porous microparticles. Research on glucan microparticles dates back to the early 1990's when Alpha-Beta successfully developed glucan carbohydrate microcapsules (Adjuvax) for targeted antigen and drug delivery. Recently, glucan particles have been used to prepare encapsulated polyplexes for DNA, siRNA, protein, and nanoparticles.

It is clear that in addition to significant biological activities, glucans also possess unique physical properties allowing them to be produced as a solution, gel, and film or particle suspension. And, when used alone or in combination with other components, glucans are useful for drug delivery applications. In some applications the glucan serves as both a carrier and the active molecule. In other applications glucan formulations have been shown useful to deliver a diverse range of drugs including small molecules, peptides and proteins, oligonucleotides, and particulate drugs. Exciting future uses of glucans for drug delivery will capitalize on the intersection of the unique immunomodulatory and physical properties of this class of polysaccharides to develop drug formulations targeting immune cells and tissues. Readers seeking more information on this subject should read an excellent review (Soto and Ostroff, 2011).

It is worthy to note that glucans have such strong and interesting effects on defense mechanisms that they have been recognized since the mid-1970 to be actively involved in plant-pathogen interactions. Glucans were found to induce the production of various defensive proteins, elicit antiviral protection in tobacco, and induce the synthesis of phytoalexins in soybeans. In addition, the defensive properties of glucans were also demonstrated in tomatoes, beans, wheat, and rice. Subsequently, a more detailed study of laminarin showed that this glucan in tobacco leaves triggered the accumulation within 48 hrs of the four families of antimicrobial pathogenesis-related proteins. In addition, the challenge of the glucan-infiltrated leaves five days after their treatment with a soft rot pathogen resulted in a strong reduction of the infection when compared to water-treated leaves (Klarzinski et al., 2000). The authors expect that glucans might become interesting alternative tools for disease control in agronomic crops.

12

GLUCAN – ALONE OR MIXED WITH OTHER SUBSTANCES?

Glucan is clearly not the only known immunomodulator in the entire world. Despite the fact that glucan, with over 12,000 published scientific papers, is the best studied and best documented natural modulator, other biologically active molecules exist. More and more manufacturers and retailers are experimenting with the preparation of various cocktails, or mixtures, of potentially bioactive powders. It is now very common to find glucan in combination with five or more ingredients, including *Echinacea, Aloe vera, Astragalus,* and *Goldenseal.*

The major problem with these elixirs is that there is absolutely no research that supports any beneficial effects. Individual extracts might have some biological effects, but they are usually either nonspecific, or their mechanisms are completely unknown. However, it still does not mean that the combination of two or more bioactive ingredients will have a greater (or even any) effect. This is particularly true in the case of complex and unpurified extracts such as the extracts of *Echinacea.* There are literally hundreds of different parts and we have absolutely no clue as to their biological activities. Some substances will have no activity, and some might stimulate, and some might inhibit the immune system. Without detailed studies, which must involve testing individual components separately as well as in combination, there is no need to play Russian roulette with our money.

Why then mix these unknown products with a well-established immunomodulator such as glucan? There are two main reasons: lack of knowledge and greed. The first one is understandable. Without deep biological knowledge, it is easy to comprehend how one can get the idea that more is better. In reality, we might have a mixture where one thing will work against the other resulting in confusing the immune system. The major reason for the absence of studies of the complex mixtures of various natural immunomodulators is financial. Biological research is expensive as it is; and the substantial testing of even one purified bioactive molecule is not cheap. When we estimate that there are usually between 10 to 100 potentially bioactive molecules in any rough extract of a plant or herb, we can imagine how expensive the purification of each one would be. And the research would come only after the purification!

The other factor I mentioned is common greed. I have nothing against selling a good product for any price the retailer decides. Considering similar qualities, it will be the market that determines if customers will pay the price or not. Unfortunately, some manufacturers/retailers are willing to put just about anything into their pills, and offer them with claims that their mixtures are better than the individual parts.

However, there are studies showing that some bioactive molecules have synergistic effects when combined with glucan. Numerous scientific studies have shown some beneficial effects when glucan was given in combination with vitamin C. The main reason why vitamin C shows synergistic effects is the fact that this vitamin has been shown to stimulate the exact same immune responses as glucan, i.e., macrophage activities, natural killer cell activity, and specific antibody formation. A mouse study showed significant healing abilities of a glucan-vitamin C combination in the treatment of infection by *Mesocestoides corti*. The treatment resulted in positive modulation of liver fibrosis and pathophysiological changes (Ditteova et al., 2003). The same group found earlier that yeast-derived glucan is a promising agent against several helminthic parasites. With respect to the liver disease, schizophyllan glucan was shown to help against an ischemia-reperfusion injury of the liver. The mechanisms of these effects are probably due to the glucan-caused decrease of the expression of immediate early genes following injury to the liver (Kukan et al., 2004).

Some of these studies were done on cattle and pigs. They showed that supplemental vitamin C and yeast cell wall glucan act as growth enhancers in newborn pigs and as immunomodulators after an endotoxin challenge subsequent to weaning (Eicher et al., 2006). However, and for some unknown reason, the majority of studies were done on fish. Numerous studies have shown the positive influence

77

of dietary glucan supplemented with vitamin C on both non-specific and specific immune responses of carp and rainbow trout (Verlhac et al., 1996). This combination is now being commercially used in fish farming.

Some studies even suggested that glucan might control blood pressure. An intravenous infusion of glucan caused a decrease in the systemic blood pressure in rats. In genetically-deficient rats with spontaneous hypertension, a diet consisting of 5 percent mushroom-derived glucan had the same effects.

Another approach in the attempt to find an ideal complement to the bioactive glucan was focused on humic acids. Humic acids represent a group of rather common high molecular weight macromolecules consisting of complex polymeric aromatic structures. They are produced by chemical and microbial degradation of organic matter originating from plants and animals. These compounds can be found in lignite, turf, soil and drinking water. Together with fulvic acids they represent certain fractions of the group of organic compounds called humic substances that are considered to be inert by some and by others to be toxic.

The effects of humic acids on the defense reaction have been known for a long time. During World War I peat extracts were used to prevent infection. Later, antimicrobial, anti-inflammatory, and antiviral properties were found. Additional studies showed stimulation of lymphocyte proliferation, stimulation of humoral immune response, and the improvement of health in farmed animals. Our own studies demonstrated significant stimulation of both cellular (phagocytosis, tumor suppression) and humoral (antibody production and cytokine secretion) branches of immune reaction (Vetvicka et al., 2010). Based on these promising results, we prepared a new, modified batch of humic acid samples and tried again. The new humic acid samples were found to be superior to the old, and, in addition to improving the synergistic effects with yeast-derived glucan on immune reactions, they also potentiated glucan's effect on wound healing. To summarize these data, our studies suggested that humic acids are biologically active immunodulators affecting both the humoral and cellular branch of immune reactions and work hand-in-hand with glucan. Carefully prepared combinations of humic acids and glucan represent another approach to increase the already potent effects of glucan. At the same time, these findings further support the study of possible synergistic effects of various combinations of natural immunomodulators.

13

Glucan and Resveratrol

Among several possible immunoactive substances already tested for possible synergy with glucan (see **Chapter 12. Glucan – Alone or Mixed with Other Substances?**), one immunomodulator stands up as potential winner – resveratrol.

Resveratrol (*trans*-3,4',5-trihydroxystilbene) is a non-flavonoid polyphenol found in various fruits and vegetables and is abundant in the skin of grapes. It exists in two isomeric trans- and cis-forms. Important to note is that, despite the fact that grapes are most commonly mentioned, resveratrol can be found in other plants, including mulberries, eucalypthus, and peanuts. Despite all published articles about resveratrol and wine, the highest concentration in nature is in knotweed. Resveratrol has major biological functions in plants, particularly as protection against fungal infection. However, resveratrol is considered to exhibit broad beneficial health effects not only in plants, but in animals and humans as well. These compounds are produced in higher plants through stress. Due to their role in plant defense, resveratrol is a naturally occurring phytoalexin ("defender of the plant").

Some of the resveratrol-producing plants are part of the human diet, including berries, peanuts, and wine grapes. Resveratrol can only be found in the skin of grapes, not in grape pulp. Due to different technological processes, the amount of resveratrol in red wine is significantly higher compared to white wine. Depending on the variety of grapes, the amount of resveratrol in wine can reach 2 to 40 µM.

Epidemiological studies demonstrated that moderate red wine consumption correlates with a lower incidence of cardiovascular diseases. This phenomenon is often called the *Mediterranean paradox or French paradox* because French people suffer less from cardiovascular diseases compared to other industrial countries, despite the fact that their diet is rather rich in saturated fats. This phenomenon was first noted in 1819 by the Irish physician Samuel Black.

About 70% of the resveratrol dose given orally as a pill is absorbed. Nevertheless, the oral bioability of free resveratrol is low because it is rapidly metabolized in the intestines and liver into several metabolites. It is, however, quite possible that these metabolites are similarly active as free resveratrol.

Intestinal absorption of resveratrol is now well documented, Fifty to seventy percent of resveratrol is absorbed in the gut; and its peak level, or its metabolites, can be observed in the blood app. 30 minutes after oral administration. It must be mentioned, however, that moderate, or even high, consumption of red wine cannot reach pharmacologically active levels of resveratrol in humans (Vitasglione et al., 2005). Consumption of resveratrol did not cause serious adverse effects in humans when administered orally at dosages up to 5 g. Readers seeking more details about the dose-dependency of resveratrol in providing health benefits should read an excellent review by Mikherjee et al. (2010).

The role of resveratrol as a possible therapeutic molecule has been considered since 1992, when it was first found in red wine. The major reasons behind the role of resveratrol in cardiovascular diseases are its promotion of vasorelaxation, suppression of artherosclerosis, decrease in serum cholesterol, and platelet aggregation.

A recent study published in the British Journal of Nutrition showed that a daily 10 milligram dose of resveratrol was associated with a reduction in insulin resistance in type-2 diabetics. In this study, researchers recruited 19 people with type-2 diabetes and randomly assigned them to receive either resveratrol supplements or a placebo for four weeks. Results showed that after four weeks of resveratrol supplementation, the participants showed a significant decrease in insulin resistance, compared to the placebo group. In terms of a potential mode of action for the polyphenol, the researchers noted that this may be related to its antioxidant activity, because oxidative stress is widely accepted to play a key role in the onset of insulin resistance. There is also the possibility that resveratrol's potential benefits are linked to its ability to activate Akt phosphorylation (Brasnyo et al., 2011).

Based on these first observations, more and more studies have lately been devoted to the effects of resveratrol. After almost two decades of intensive research, it is now well-established that resveratrol is biologically a very active molecule and belongs in the top immunomodulators. In addition to various biochemical, biological and pharmacological activities, resveratrol has been found to exhibit numerous immunomodulatory activities, such as suppression of lymphocyte proliferation, changes in cell-mediated cytotoxicity, cytokine production, or induction of apoptosis.

ANGIOGENESIS

Angiogenesis is the formation of new blood vessels from existing ones and is required for the growth of tumors beyond a diameter of 1 mm. It is clear that angiogenesis is an extremely important feature in cancer growth, and its spread. Is it well-established that the number of vessels in cancer is closely related with the prognosis of patients. Resveratrol inhibits vascular endothelial growth factor-induced angiogenesis by the inhibition of specific enzymes. For details regarding the role of resveratrol against angiogenesis, go to (Kraft et al., 2009).

APOPTOSIS

Apoptosis is the process of programmed cell death that may occur in multicellular organisms. Biochemical events lead to characteristic cell changes and death. These changes include blebbing, loss of cell membrane asymmetry and attachment, cell shrinkage, nuclear fragmentation, and chromosomal DNA fragmentation. Unlike necrosis, apoptosis produces cell fragments that surrounding cells are able to engulf and quickly remove before the contents of the cell can spill out onto surrounding cells and cause damage. As apoptosis might be involved in possible cancer suppression and represents a possible therapeutical tool, it is not surprising that resveratrol was also tested as possible apoptotic agent.

Detailed studies showed that gastric adenocarcinoma cells responded to resveratrol with inhibition of DNA synthesis, cell cycle arrest, suppressed proliferation, and induction of apoptosis (Riles et al., 2006). Further studies revealed that resveratrol engages selective apoptotic signals that are cell-specific—most of all p53 protein. The findings, that in some adenocarcinoma cells resveratrol upregulated p53 and downregulated survivin, whereas in other cells stimulated caspase 3 and cytochrome C oxidase activities, indicate that even within a specific disease resveratrol can engage alternate apoptotic targets. Details regarding the role of resveratrol in apoptosis can be found in (Kraft et al., 2009).

HEART DISEASES

In addition to other biological activities, resveratrol   has been found to exhibit cardio protective effects related to its antioxidant activity. This original observation led to the subsequent studies of its effects on so-called preconditioning (i.e. brief period of ischemia resulting in protection against heart injury in subsequent ischemic insult). An entire series of detailed, and highly complicated studies, revealed that resveratrol caused preconditioning effects, which are mediated by nitric oxide or nitric oxide synthetase (Hattori et al., 2002). In addition, resveratrol enhanced myocardial angiogenesis.

ANTI-INFLAMMATORY EFFECTS

Similar to other biological effects of resveratrol, the anti-inflammatory effects are multifactorial; this means that the compound affects several aspects occurring during inflammation (readers seeking more information should read Kraft et al., 2009). Some of the anti-inflammatory effects are manifested during cardiovascular protection, most notably in reducing myocardial infarct sizes and improving post ischemic ventricular function.   Recent studies further support the hypothesis that the cardio protective effects of resveratrol are mediated via its anti-inflammatory action (Csiszar et al., 2006).

EFFECTS ON CANCER

Resveratrol has the ability to affect critical signaling mechanisms in tumors. Currently, it is well established that many factors participate in the pathogenesis of cancers. However, the exact mechanisms are currently undefined. The first reports describing the role of resveratrol in the fight against cancer were published in 1997 (Jang et al., 1997). These studies showed that resveratrol is able to exert cancer therapeutic activity at all three major stages of carcinogenesis, i.e. anti-initiation activity, anti-promotion activity, and anti-progression activity. It is no surprise that these first findings started an immense interest in resveratrol and its healing properties. Studies demonstrated that resveratrol inhibited free radical formation and inhibited mutagenic response induced by treatment with anthracene.

Animal studies showed that resveratrol inhibited tumor growth by as much as 80%. One of these studies focused on resveratrol in the development of intestinal tumors. Using a model of mice genetically predisposed to developing intestinal tumors, the study showed that oral administration of resveratrol initiated a dramatic decrease in the number of tumors in the small intestine and completely suppressed tumor formation in the colon. By analyzing differential gene expression patterns in the cells of the small intestine, the investigators were able to show that resveratrol downregulated a panel of genes directly involved in the progression of tumorigenesis, and upregulated several genes controlling genome stability, and cellular differentiation, and favored the activation of antitumoral defense mechanisms.

Further detailed studies revealed that resveratrol has strong effects on several genes involved in both the development of cancer and the fight against cancer. On one hand, resveratrol stimulated genes involved in apoptosis, and on the other hand, downregulated expression of antiapoptotic proteins. Resveratrol is also a sensitizer of tumor necrosis factor-related apoptosis. These findings provide further evidence that resveratrol can be considered a versatile chemopreventive agent. An interesting study using a myeloma model showed a strong potential of resveratrol as a drug for treating multiple myeloma.

In summary, resveratrol was found to act as an antioxidant and antimutagen and to induce phase II drug-metabolizing enzymes. It mediated anti-inflammatory effects and inhibited cyclooxygenase and hyperoxidase functions (antipromotion activity) and induced human promyelocytic leukemia cell differentiation (antiprogression activity). In addition, it inhibited the development of preneoplastic lesions in carcinogen-treated mouse mammary glands in culture and inhibited tumorigenesis in a mouse skin cancer model. During all these studies, no toxic effects of resveratrol were observed. These data suggest that resveratrol, as a constituent of the normal human diet or given as a supplement, merits detailed investigation as a potential cancer chemopreventive agent in humans.

INTERACTION WITH OTHER DRUGS

One cannot overlook the possibility of interaction of resveratrol (or any other biologically active molecule, from drugs to natural biological modulators) with other substances. In the case of resveratrol, only a limited number of possible negative interactions were observed. One of those is the enhancement of the effects of resveratrol on NO activity by ethanol which is another stimulation of anti-platelet effects of prostaglandins.

On the other hand, resveratrol has been found to have positive synergistic reaction with several drugs, mostly with chemotherapeutic drugs. One of those is cisplatin—a common cytostatic drug used in the treatment of numerous tumors. However, high nephrotoxicity often limits its use. Resveratrol blocks cisplatin-induced structural and functional renal changes which are most likely caused by reducing free radicals (Do Amaral et al., 2007). Resveratrol is similarly effective in the case of doxorubicin which also has strong negative side-effects. Resveratrol improved the efficiency of doxorubicin, allowing us to use a lower, safer dose of this chemotherapeutic agent. Good synergistic effects were also observed in the case of paclitexel, where resveratrol was shown to sensitize chemoresistant tumor cells to the chemotherapeutic treatment (Gupta et al., 2011). Reader seeking more details on role of resveratrol in cancer prevention should read an excellent review written by Bhat and Pezzuto (2002).

Additional studies focused on the combination of resveratrol with anticancer drugs (such as cycloheximide, busulfan, gemcitabine and paclitaxel) in the human leukemic cell line HL60 and its multidrug–resistant subline. Cell cycle analysis showed that treatment with resveratrol resulted in cell cycle arrest of both variants. The ability of resveratrol to influence one or more critical pathological factors raises the possibility of its use in combination with anticancer drugs that lead to additive or synergistic therapeutic effects.

IMMUNOLOGICAL EFFECTS

With all the biological effects mentioned above, it is not surprising that resveratrol was also tested as an immune reaction-enhancing component. In the immune system, resveratrol was reported to inhibit the production of nitric oxide and tumor necrosis factor (TNF-α) from macrophages. Resveratrol has also been tested as a potential anti-inflammatory agent. Using a model of carrageenan-induced inflammation in rats, resveratrol was found to significantly reduce edema both in acute phase (3 to 7 hours) and in the chronic phase (24 to 144 hours).The edema-suppressing activity of resveratrol was comparable to that of indomethacin. A study using human promonocytes showed enhancement of phagocytosis (Bertelli et al., 1999).

As the optimal dose of resveratrol is often questioned, a particularly interesting study observed the effects of a low dose of resveratrol on immune response. Concanavalin A and/ or *Staphylococcus aureus* were used to induce the activation of T lymphocytes, antigen presenting cells, and cytokine production. The results showed that resveratrol induced lymphocyte proliferation and IL-2 production, stimulated production of IL-12 and interferon gamma, and also inhibited production of IL-10. Higher doses of resveratrol promoted delayed type hypersensitivity. Lymphocyte subtypes were not significantly affected. The conclusion of this study was clear—low doses of resveratrol enhanced cell-mediated immune responses by activating macrophages and promoting Th1 cytokine production (Falchetti et al., 2001).

However, resveratrol acts not only on the cellular level but also on the genomic level. Numerous studies showed that resveratrol influenced the expression of NF-κB, TNF-α, interleukin-1, or interleukin-6.

Our group also decided to test how resveratrol can influence an immune system. Our study was based on a recent observation showing that seaweed-derived glucan elicited defense responses in grapevine and induced protection against Botrytis cinerea and *Plasmopara viticola* via the induction of production of two phytoalexins including resveratrol (Aziz et al., 2003). This led us to evaluate the possible synergetic effects of glucan and resveratrol on immune reactions.

Our   study showed that both glucan and resveratrol stimulated the phagocytosis of blood leukocytes, caused the increased expression of some membrane markers (such as CD4) on spleen cells and showed higher restoration of spleen recovery after experimentally induced leucopenia. In all cases, strong synergetic effects were observed. When we measured the effects of these substances on the expression level of some important genes (such as NF-kB2, Cdc42 and Bcl-2) in breast cancer cells, the up-regulation of Cdc42 expression was evident only with the use of both immunomodulators (glucans and resveratrol) in combination (Vetvicka et al., 2007 b). The up-regulation of NF-κB2 gene expression is considered significant since the members of this family are important regulators of cell cycle progression, cell survival, cell adhesion/angiogenesis, invasion and inflammatory responses. Studies have shown the positive role of the NF-κB family proteins in regulating the expression of adhesion molecules and angiogenic factors that are known to increase the invasion and metastasis of cancer cells.

Figure 19 demonstrates the strong synergetic effects of a glucan—vitamin C-resveratrol combination on the phagocytosis of peripheral blood neutrophils. These data were confirmed by three different commercial samples of resveratrol. Several conclusions can be reached from these results: 1) glucan works better with resveratrol than alone; 2) the combination of all three compounds consistently showed the strongest effects. Additional studies confirmed that glucan and resveratrol affected cell-mediated immunity even more when used together with vitamin C.

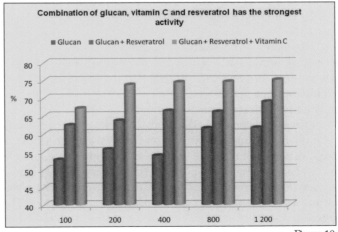

Figure 19

86

Similar data were obtained in experiments measuring the effects of glucan and resveratrol on their own or a glucan-resveratrol complex on the production of important cytokines. The production of all three tested cytokines (IL-1, IL-6 and tumor necrosis factor alpha) was significantly increased. Figure 20 clearly indicates that, whereas glucan alone showed no stimulating activity and resveratrol alone had some activity, only in the case of IL-1 and IL-6, did the simultaneous treatment with glucan and resveratrol result in a strong secretion of all three tested cytokines.

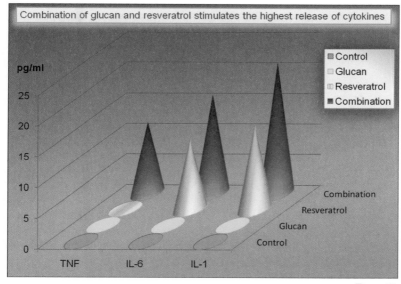

Figure 20

Similar to cyclophosphamide, cancerostatic drug 5-fluorouracil is well known for its significant depression of the immune system. We evaluated the effects of orally-given glucan and resveratrol on fluorouracil-induced leucopenia. The data obtained in this study showed that, whereas glucan strongly increases the recovery of bone marrow, resveratrol alone showed significant stimulation only after 10 days of application. When combined, glucan and resveratrol had small, albeit insignificant, synergetic effects. Things looked quite different in the spleen, though. Even if glucan was again significantly more active than resveratrol, its effects showed a substantially higher restoration of spleen cellularity from day 4. Combined glucan-resveratrol substances showed a stronger synergetic effect that was significant from day 11 (Vetvicka et al., 2007).

Next, we focused on the possible effects of resveratrol-glucan complex on humoral branch of immune reactions. We used a well-established model situation of a specific antibody response against ovalbumin. Data shown in Figure 21 demonstrate that resveratrol alone has only limited effects on antibody response, and, based on current knowledge of its abilities, was not entirely surprising. When we tested glucan alone, we found that the level of stimulation corresponded to previously published data (Vetvicka and Vetvickova, 2011). However, the resveratrol-glucan combination stimulated antibody formation almost to the level obtained with Freud's adjuvant.

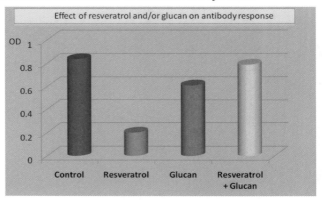

Figure 21

Since the effects of resveratrol on apoptosis are well established, we compared the synergistic effects of the resveratrol-glucan combination. Results summarized in Figure 22 showed that whereas glucan alone has only very limited effects, resveratrol alone increased apoptosis almost three times. When used together, these effects were four times higher.

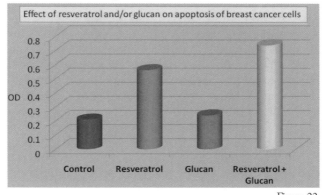

Figure 22

Resveratrol is reported to be a strong scavenger of free radicals. Several studies suggested that this compound is an active anti-oxidant *in vivo*, due mostly to the stimulation of nitric oxide formation and by maintaining the concentration of intracellular antioxidants present in biological systems. However, direct studies measuring the anti-oxidative effects *in vitro* did not demonstrate any activity. On the other hand, glucan is considered to be a strong antioxidant. However, the available data are based on the in vitro experiments. It was therefore of interest to compare the scavenging activity of both supplements. In our study we used a 1,1'-diphenyl-2-pictylhydrazyl model, and again evaluated the possible anti-oxidative effects of resveratrol, glucan and a combination of both molecules. As you can see from the Figure 23, the combined resveratrol-glucan substance was clearly the most active. These data also confirmed the previous results—where glucan either helps resveratrol to reach higher activity or to be active in areas where resveratrol alone is not active.

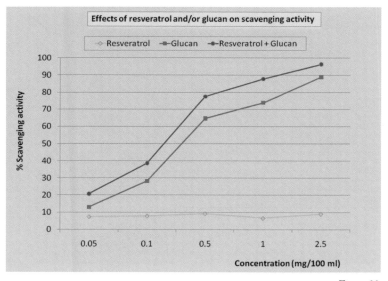

Figure 23

Clearly, enough data has been accumulated in scientific literature to allow us to conclude that resveratrol acts as an antioxidant and anti-mutagen and induces phase II drug-metabolizing enzymes, and mediates anti-inflammatory effects. In addition, it inhibits the development of preneoplastic lesions in carcinogen-treated mouse mammary glands in culture and inhibited tumorigenesis in a mouse skin cancer model. During all these studies, no toxic effects of resveratrol have been observed. These data suggest that resveratrol as a constituent of the normal human diet or given as a supplement, merits detailed investigations as a potential cancer chemopreventive agent in humans. In addition, resveratrol has been found to have anti-aging activities, stimulates the immune system, and can exhibit cardio protective effects. To make a long story short, resveratrol exhibits significant beneficial health effects. And to make a good short story even better, the addition of glucan makes resveratrol even more potent.

14

WHICH GLUCAN TO CHOOSE?

Glucans can be isolated from numerous sources including yeast, mushrooms, and grain. In fact, glucans can be, and have been, isolated from almost every species of mushroom and yeast. In addition, several additional, more exotic sources of glucan have been described, including lilies and protist *Euglena gracilis*. The decision is based primarily on the cultural history of an individual country. The Far East has a long history of mushrooms as an old folks remedy and therefore, choosed mostly mushroom-derived glucans, whereas the United States and Europe, having a long tradition of eating bread and drinking beer, have a surplus of yeast and therefore focused on the use of yeast-derived glucans. Any conclusion is made even more difficult by the fact that glucan as a supplement is consumed virtually in every civilized country of the world. Every reseller will stress the quality of his product, but in order to make sure we will get the best product and not waste money on some inferior (or in the worst scenario complete-ly untested) glucan. Therefore, it is imperative to do our own research of available literature.

This takes us back to the original question: out of dozens, if not hundreds, of individual glucans on the current market, which is the best one? Which one has superior biological and/or immunologi-cal properties? For years, there was a controversy between the no-tions that water-insoluble glucans show only little biological activity, whereas soluble glucans are highly active. The scientific literature on glucan running close to 12,000 papers, is the most extensive of all the natural immunomodulators studied. It is only recently that a number of papers (many of which are mentioned in this book) have shown that the question of bioactivity is more to do with the purity, and the physicochemical properties, than with solubility (Hong et al., 2003).

The original studies regarding the effects of glucan on the immune system were focused on mice. Subsequent studies demonstrated that β-glucan has a strong immunostimulating activity in a wide variety of other species, including earthworms, shrimp, fish, chicken, rats, rabbits, guinea pigs, sheep, pigs, cattle and humans. Based on these results, it has been concluded that glucan represents a type of immunostimulant that is active over the broadest spectrum of biological species and represents one of the first immunostimulants active across the evolutionary spectrum. Glucan is a biologically active polysaccharide that can be considered to be an evolutionary, yet extremely old, stimulant of all types of defense reaction. This idea was further confirmed by experiments showing that glucan also plays a role in the defense of plants.

In addition, glucan exists either unchanged or chemically modified. Because the *carbohymethylation* increased the solubility without changing the biological properties, *carboxymethyl glucan* has been used as a substitute for soluble glucans for decades. Similarly, some studies successfully used glucan phosphate (Li et al., 2004). Some other studies suggested that sulfation might increase the biological activities or even potentiate some new effects. As naturally sulfated polysaccharides generally have antiviral properties, it is not surprising that experimental sulfated glucans were also successfully used for their antiviral activity. We used a sulfated seaweed-derived glucan PS3 with a degree of sulfation of 2.4. Previous studies on plant defense showed a stronger effect than with normal seaweed glucan, suggesting that chemical sulfation might increase its properties. Our observations found that this sulfated glucan had almost identical properties as the normal ones (Vetvicka et al., 2008 b). Several other groups later confirmed our results.

Because of the lack of comprehensive reviews comparing the biological effects of glucans isolated from various sources, and despite extensive investigations, no final conclusion has been reached. Therefore, there is not a single research paper willing to state that one source of glucan is better than another. In addition, numerous concentrations and routes of administration have been tested, including intraperitoneal, subcutaneous, and intravenous applications. For decades, oral treatment with glucan has been on the periphery of interest, despite the fact that it represents the most convenient route. In the last decade, however, a renewed interest in human applications brought about important studies of the orally-administered glucan. In the past decades, our research group focused on comparing several of the most successful commercially and easily available glucans on the market. We chose glucans widely sold and available around the globe, representing grain, mushroom, and yeast-derived glucans in both soluble and insoluble forms. The results of these comparisons were published in numerous peer-reviewed scientific journals (Vetvicka and Vetvickova, 2005, 2007 a, b, 2008, Vetvicka et al., 2008 a, Vetvicka and Vetvickova, 2011) and are summarized throughout this book.

From these data and from careful comparison of other studies, it is evident that all glucans are not created equal. Yes, both the Ferrari and Ford Escort are cars and both will get us to our destination eventually. But, the level of comfort and pleasure will be substantially different. The real importance in deciding which beta glucan to buy and to consume lies in finding a respectable company. Clearly, every retailer claims that his glucan is the best, and that the competitors are selling inferior products and the question arises once again.

The most important aspect is to either manufacture, or purchase, glucan from a solid source and then control the quantity, the purity, and the biological activity. Beware of middle-men and, if possible, buy directly from manufacturers or from resellers with clearly demonstrated connection to the manufacturers. In order to obtain the relevant information, it is imperative to get through the smoke and mirrors of unsupported or sometimes pseudoscientific claims about the glucan offered. There are some telltale signs that should immediately raise a red flag. One of them is a full database of glucan papers without any direct proof that the research was done using the beta glucan which the company is actually selling. It is very easy to find dozens of papers describing the significant effect of glucan, but if the particular study is not done for the glucan of interest, it is not relevant to the claim. Similarly, some retailers are hiding behind patented manufacturing processes, which are legitimate points, but rarely support the claim about the quality of the product, since they describe the manufacturing process and not the biological activity. It is also important to pick a manufacturer/reseller, which specialize on supplements only, and, if possible, with a limited range of products. I do not have to explain that the chance of getting a well-documented, and well-tested glucan is much higher with somebody specializing in few natural supplements than with a seller offering dozens of different supplements manufactured by dozens of different companies. If he offers other products beyond the supplements, I would sincerely recommend using an entirely different company.

Some retailers and even manufacturers are using pseudoscientific claims, such as the advantage of extra-small size, suggesting that cells involved in internalization and subsequent transfer of glucan, such as dendritic cells, macrophages and neutrophils, have "receptors for the beta glucan approximately 1-2 microns in size." It is clear that these salesmen do not realize what they are talking about, as macrophages have glucan receptors the size of only several molecules. Yet, they are able to phagocytose material of 20% their own size. It is apparent that, in the case of macrophages and phagocytosis, size really does not matter. Numerous scientific studies clearly demonstrated not only this, but also that, glucan can be soluble or insoluble and still be highly active. Similarly, some companies are selling micronized glucans that are often accompanied by claims that they are more bioavailable. It might be true. It is unfortunate that, thus far (at the time of going to press), there is no published scientific evidence to support this statement.

The question of purification can be also confusing. In an ideal world, every glucan on the market would be 100% pure. In reality, it is commercially almost impossible to achieve this purity and still keep the price in the reasonable range. The level of purity is still important, but, as always, the devil is in the details. The definition of glucan content varies according to the manufacturers' methods. Some companies calculate glucan content by the percentage of carbohydrates, hexose, or glucose. That would work fine in the case of 100% glucan, but not if the sample contains glucan plus other carbohydrates such as mannoproteins or even glucose. Therefore, it is important to look for a product that guarantees the beta glucan content. We tried to compare the glucan content among different commercial glucans, and the results were quite surprising (Figure 24). One cannot always get fair value for one's money. When we compared the retail price of the same glucans, the results were mind boggling (Figure 25).

Another common way used to turn glucan into a higher quality product than it actually is, involves vague claims that the samples were tested by some medical school. Without a direct link or even excerpts from

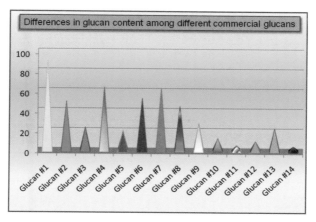

Figure 24

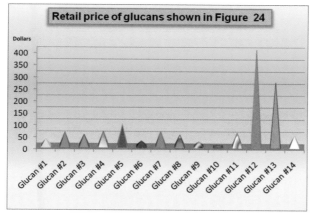

Figure 25

the independent scientific study that described the biological activity of the glucan in question, similar claims are worthless. *The Journal of the American Nutraceutical Association* has recently published several important articles focusing on the comparison of various glucans (Vetvicka and Vetvickova, 2005, 2007 a, 2008 a). Readers seeking the best possible glucan should carefully read these and similar research articles mentioned in this book.

It is time to return once more to the original question: which beta-glucan to choose? Well, it might be easier than you think. It is extremely important to choose a manufacturer, or retailer closely connected to the manufacturer, who has invested in careful isolation and purification, but also into research. In doing only this, one can be certain that the beta glucan purchased will be doing its part in helping the immune system to fight intruders. If your immune system could thank you, it would.

15

GLUCAN IN ANIMAL TREATMENT

As previously mentioned, glucan is one of the few bioactive molecules which successfully crossed the species barrier, i.e., that are similarly active in humans and such "lowly" animals such as shrimp. For many years, all the research on glucan was conducted entirely with animals. Therefore, it is not surprising that glucans were and continue to be used in a variety of animal species.

Animal experiments are very informative, since they were carried out under strictly controlled conditions. Unlike in human studies, the results were never influenced by subjective reporting. Glucan products are currently in practical use as an adjuvant in vaccines and in feed for shrimp, fish, and numerous warm-blooded animals, including pets.

Glucans from baker's yeast and other bacterial, fungal, and plant sources have been recognized as potent immunomodulators in different fish species. Recent breakthroughs in the industrial manufacturing of yeast glucans allow for its use as an affordable dietary supplement for aquaculture and pet fish. Several Norwegian groups have studied the effects of glucan on fish for decades. The production of fish larvae is often hampered by high mortality rates. It is believed that most of this economic loss is due to infectious diseases and reaches as much as 10% in the Western European aquaculture sector. The development of strategies to control the pathogen load and prophylactic measures must be addressed further to realize the economic potential of marine fish larvae, thus improving the overall production of adult fish. The innate defense includes both humoral and cellular mechanisms such as the complement system and the process played by granulocytes and macrophages. Natural substances such as glucans were found to directly initiate the activation of the innate defense mechanisms acting on receptors and triggering intracellular gene activation, resulting in the production of anti-microbial molecules. Glucan acts as a true adjuvant in fish, enhancing their antibody production. An even more important fact is the similarly high effects observed in injected and orally-given material.

The use of immunostimulants as dietary supplements improves the innate defense of animals, providing them resistance to pathogens during periods of high stress, such as grading, reproduction, sea transfer, and vaccination. The immunomodulation of larval fish has been proposed as a tool for improving larval survival by increasing the innate responses of the developing animals until their adaptive immune response is sufficiently developed to mount an effective response to the pathogen (Dalmo et al., 1997).

As previously mentioned, most of these studies originated in Norway where a booming industry preparing glucan for commercial aquaculture exists. It is not surprising that most of the farmed salmon currently on the market has been treated with glucan. A large series of studies was performed by the Norwegian Institute of Fisheries and Aquaculture. They showed that the addition of glucan to the feed significantly lowered mortality caused by the infection with *Piscirickettsia salmonis*. A combined preparation of glucan and vitamin C showed even stronger effects (see **Chapter 11: Glucan – Alone or Mixed with Other Substances?**). The addition of glucan to fish food resulted in an anti-stress diet, greatly reducing problems with vibriosis and furunculosis, which has plagued fish farmers for decades. Likewise, glucan in food helped to protect the fish against one of the major disease problems in salmon farming: Infectious Pancreatic Necrosis.

Antibiotics are commonly incorporated into the diets of various farmed animals including broiler chickens, fish, or pigs. The reason for their use in sub-therapeutic levels is to improve digestion, growth, and feed efficiency. However, in recent years, there has been a tremendous increase in bacterial resistance to antibiotics, in both human and animal populations. This caused increased consumer concern for drug residues in meat. As of January 1, 2006, the European Union imposed a complete ban on the use of antibiotics in animal feed as growth promoters. Other countries, including the United States, are considering similar measures.

Glucans, particularly glucans derived from yeasts, are promising alternatives to antibiotics, as they have been shown to improve growth performance and stimulate the immune system in immature broilers. Results of the study performed by Atlantic Poultry Research Institute showed that glucan may effectively replace *virginiamycin* as a growth promoter for roosters or chickens up to 38 days of age.

Based on the above-mentioned studies, it is not surprising that glucan is more frequently tested and used in commercial animal feed. Canadian universities focused on grain glucans and developed the glucan concentrate *Viscofiber*, which is now commercially available for use in functional food and dietary supplement. *Progressive Bioactives* prepared a glucan additive (PROVALE™) that is being tested as an additive in nursing pigs and pregnant sows. Similarly good results were obtained using *Agrastim™* in cattle. In an *in vitro* study done at the University of West Virginia, *Agrastim™* was mixed with rumen fluid, and the disappearance of dry matter was measured. Orally-given *Agrastim™* (beta-glucan) clears the rumen within two hours, thus reaching the Peyer's patches in the gut. A follow-up study showed that *Agrastim™* added to the replacement milk significantly improved the survival of calves.

The same glucan was also tested in the treatment of mastitis. Mastitis is an inflammatory reaction of udder tissue most commonly associated with bacterial infection. It is the most common and, at the same time, the most costly dairy cattle disease. One of the signs of mastitis is an increase in blood proteins and white blood cells in the mammary tissue, then subsequently in the milk. Since approximately 40 percent of dairy herds are affected, the costs are astronomical. The Addition of *Agrastim* to the feed at a dose of 2 g/day resulted in a drastic decrease of the blood cell content in the milk, showing that glucan aids in maintaining low blood cell counts in milk by controlling subclinical mastitis and other infections of the udders.

Yeast glucan is able to absorb several mycotoxins (such as zearalenon, aflatoxin B1, deoxynivalenol, ochratoxin A, patulin) and heavy metals, probably through physical adsorption, hydrogen bonds, and van der Waals forces; these ß-glucan effects are important particularly for livestock. Krakowski and his group studied the effects of glucan administration on cellular and humoral immunity in foals, in the neonatal and postnatal period, and found strong stimulation of cellular immunological mechanisms (Krakowski et al., 1999).

Similarly, the addition of the yeast-derived glucan into the food of weaned piglets was found to be helpful in increasing the proliferation of blood lymphocytes, the effectiveness of swine fever antibodies and increased production of IL-6 and IL-8 (Wang et al., 2007). In other experiments, piglets receiving glucan showed improved appetites, which, together with the greater ileal digest, bacterial richness and diversity and reduced colonic ammonia, indicates a healthier gastrointestinal environment. Another commercially important animal with demonstrated positive effects of glucan is lamb, where supplementation of the diet with yeast-derived glucan showed significantly higher humoral and cellular immunity (Wojcik et al., 2010).

Glucan was also shown to boost the health of commercially farmed chickens. Glucan-improved food helped enhance the phagocytic and bactericidal capacity of macrophages in chicks, which resulted in blocking the entrance of harmful *Salmonella enterica* (Chen et al., 2008). Similarly, positive effects on the health of farmed pigeons and ducks have been found.

From the above mentioned information, a simple conclusion can be reached: glucan has significant potential in the activation of the relevant defense reactions in all types of farmed animals, including shrimp, fish, and livestock. Glucan will help to achieve more healthy animals and significantly lower the use of vaccines and antibiotics, still used in commercial farming. As glucan is a natural molecule isolated from natural sources, this information might be particularly important to bio farming.

16

POSSIBLE USE OF GLUCANS

What are the major benefits of taking beta glucan? This nutrient benefits anyone who wants to be healthier, live longer, deal with the stress of modern society, be less allergenic, speed up healing and resist the infectious microbes, bacteria and viruses that seem to be everywhere. As you have learned, the major reason to take glucan is to enhance your immune system. Based on current literature, there are several possible ways to use glucan. Generally, these possibilities were demonstrated in various animal models including mice, rats, guinea pigs, rabbits, horses, chickens, fish, sheep and cattle, and can most probably be extrapolated to humans.

Glucan has been shown to have a significant role in the following biological processes:

- Anti-infection immunity
- Anti-cancer immunity
- Wound healing
- Support of myelopoiesis
- Non-specific stimulation of defense reactions
- Lowering of cholesterol
- Anti-irradiation protection
- Adjuvant effects
- Stress

Glucan has a particularly strong role in the activation of the immune system, including increased activity of macrophages, neutrophils and NK cells, increased phagocytosis, increased production of reactive oxygen species and free radicals, stimulated production of crucial cytokines such as IL-1, IL-2, IL-6, TFN-α, increased natural cytotoxicity, increased wound healing, increased anti-tumor activities, protection against irradiation and chemotherapy, and stimulation of cell formation in bone marrow.

WHO WILL BENEFIT?

From all the information mentioned so far, it is clear that everybody will benefit from glucan. However, the groups of people that are considered to benefit most from taking glucan as a supplement are:

- People with impaired immunity from any cause
- People with high occurrence of infection or undergoing chemotherapy and/or irradiation
- People over 55 years of age, when the natural aging results in slowing the immune system
- People under extensive or long-term stress
- People who exercise excessively or who are under strong physical stress (professional athletes, manual workers)
- People with a high risk of atherosclerosis due to a high cholesterol level
- People living in an environment full of UV irradiation, electro-magnetic fields, (living close to high current wires, heavy cell phone users, patients undergoing radiation therapy)
- People living close to a nuclear power station
- People consuming a high amount of fish (potential mercury poisoning)

Until recently, glucan was usually suggested for adults only. This notion was not based on any observed problems with glucan in children, but on the general way of thinking that only adults and the elderly really need to boost their immune system. This is fundamentally wrong, because the immune system of a youngster (in any biological species) is immature, still learning, and reaching its full capacity at approximately 15 to 18 years of age (in humans). From that point, the activity of the immune system gradually decreases. The truth is that, until recently, glucan was tested entirely on the adult population of animals. However, our group evaluated the immunological effects of three different glucans on young and adult mice. We found that glucan is similarly active in both age groups (Vetvicka et al., 2008 a). Therefore, when needed, glucan can be used in children as well. It is, however, important to remember that the dose has to be significantly lowered.

Thus far, we have discussed the option to use glucan as an oral supplement. However, recent trends lean towards the idea of adding glucan directly to our food. Whereas cereal producing companies are pushing the benefits of glucan present in their products, it is not clear if they are talking about glucan naturally present in their cereal, or if they are really adding a purified material to the flakes. Recently, Biothera has cooperated with several companies in adding yeast-derived glucan to their products. As for example, Good Cacao uses organic ingredients (including glucan) to create a rich dark chocolate base, and there are ready-to-serve soups enriched with glucan (Lassonde Specialities). There is nothing wrong with adding glucan directly to the diet. However, since we do not consume larger amounts of chocolate on a daily bases or we do not eat one of the three Canton soups every day, it would be hard to achieve a regular daily supply of glucan. So far, none of the research papers suggested that taking glucan only from time-to-time would be sufficient to achieve full benefits of glucan. Therefore, I cannot recommend the use of this type of food as a substitute of regular consumption of glucan supplements.

OPTIMUM DOSAGE

It is important to note that the overwhelming majority of glucan research was done on animals where the doses often used do not directly correlate with doses used in humans. The major reason for this is the different sensitivity of various animal species to the biological effects of individual substances. For example, it is well known that the amount of material necessary to develop an antibody response in mice and rabbits is almost the same, despite the huge differences in body weight.

Nevertheless, after thousands and thousands of research studies, one can apply these findings to humans. It is true that all substances acting as immunostimulants do not show a linear dose response, i.e., the level of stimulation is not a direct reflection of doses. Instead, the maximum activity is within an intermediate concentration. In addition, one cannot always use common sense that might suggest that two pills of 50% glucan will have the same effect as one pill of 100% glucan. We evaluated these claims in detail, and our research clearly showed that the final results depend more on the initial quality of the individual glucan than on its concentration. We compared the effects of several commercially available glucans on the phagocytosis of cells. We picked glucans isolated from yeasts, mushrooms and barley: seven yeast-derived glucans, two mushroom-derived glucans, and two barley-derived glucans. The differences observed in our experiments were stunning. On the one hand, some glucans elevated the phagocytic activity of cells up to almost 60%. On the other hand, some commercial glucans had no activity at all (Figure 26).

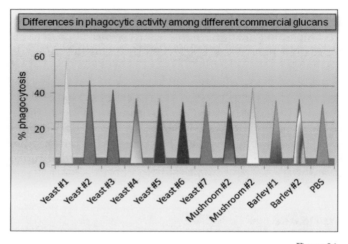

Figure 26

Some glucans need several thousand times higher dosages to achieve the activity of the top glucans. Others will never reach the top quality regardless of the dose (Figure 27). In some cases, the bioactivity of glucan can gradually decline after reaching the optimum dosage levels.

There is no absolute rule about the exact dose of glucan to be taken to maximize the activation of our defense mechanisms. However, from the vast amount of studies on animals ranging from mice to horses, we can conclude that the optimum oral dose of the good quality glucan is in the range of 0.1 to 1.5 grams per 200 pounds body weight per day. The dose, of course, depends on the overall activity

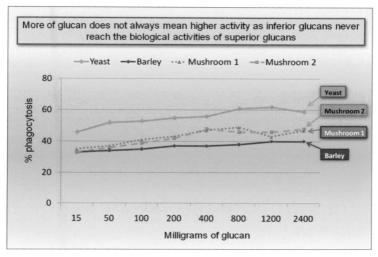

Figure 27

of the immune system and on the expected use, i.e., the lower end of the dosage is used for prophylactic or maintenance stimulation of the immunity, whereas the high end is used when our body really needs a fast boost of defense reactions, i.e., when the first symptoms appear and during the illness. A doctor should be consulted to better evaluate the individual medical conditions and corresponding dosages.

17

POSSIBLE SIDE EFFECTS

Our knowledge of possible negative effects of various ß-glucans is limited. In fact, considering almost 12,000 scientific papers, only very few reports deal with potential side effects.

Particulate glucan applied parentally was reported to cause some granuloma formation and local inflammation (Di Luzio et al., 1979). These problems, however, occurred only when glucan was administered intravenously (Zekovic et al., 2005). Inhalation of intact cells or cellular detritus of fungi or yeasts—ingredients of home dust or different agricultural and industrial dusts (Sato and Sano, 2003)—induces the so-called "syndrome of toxic organic dust", which is characterized by lung reactions that include pneumonia, cough and chronic bronchitis (Alwis et al., 1999), rhinitis, headache and irritation of the eyes and throat. It is possible that glucans might serve as causal factors in these complaints (DeLuca et al., 1992, Fogelmark et al., 1994). However, the use of glucan as a natural immunomodulator does not include inhalation. All the problems described above are based on long-term inhalation of very high doses. Some studies even used direct injection of glucans into the lung tissue. Clearly, these treatments are not physiological and reflect very little of a normal situation. Even using these drastic conditions, the results of these studies are controversial and repeatedly showed no negative effects.

The Swedish Biofact Environmental Health Research Center has for years been trying to evaluate the potential effects of inhaled glucans. Using both animal and human studies, there is still no clear-cut conclusion, partly because individual studies employed different glucans. Extractions of glucan from the fungal or mushroom wall might affect the chemical characteristics and hence the biological activity. Most of the studies using the glucan levels corresponding to the levels found in the environment, found no inflammatory or other responses. Only when the doses used reached extreme levels, inflammation of lungs, probably due to the stimulation of cytokine secretion, was described. The research center proposed a model situation where the high level of glucan in the environment might result in stimulation of body defenses and subsequent increased risk of some diseases. However, it is important to understand that this model and these conclusions **do not mean** that there is any risk in using glucan as a supplement. These conclusions only point out that there might be a significant risk to live in an environment full of mold.

One potential question has been raised several times during the past decades – are there lethal side effects when glucan is combined with indomethacin? Several reports described that β-glucan-induced inflammatory processes could competitively interact with simultaneously administered non-steroidal anti-inflammatory drugs (such as indomethacin). A lethal toxicity due to septic shock, elicited by a sequential administration of ß-glucan and a non-steroidal anti-inflammatory drug indomethacin, was described in mice (Takahashi et al., 2001, Yoshiba et al., 2001). The authors believe that the sepsis was induced by damaging mucosal membrane and resulting transmigration of intestinal bacteria into the blood stream. In addition, these effects can be reversed by antibiotics such as fradiomycin and polymyxin B (Nameda et al., 2007 a, b). The effects strongly vary according to the laboratory strain of mice, suggesting that some particular combination of genes might make the animals susceptible. More details can be found here (Ohno, 2011). Unfortunately, all these reports are from one laboratory and, despite the fact that these effects were repeatedly described for more than 12 years, they were never independently confirmed.

Due to the lack of independent confirmation, we decided to perform a study that would either confirm or reject the current theory regarding the negative association of glucan and non-steroidal anti-inflammatory drugs. In addition, as individual glucan often differ in their biological properties, we wanted to establish that our findings have a general meaning. Therefore, we used four different types of glucan, differing in source, solubility and activity, and combined them with indomethacin. We also tested the various routes of administration, i.e., both injected and orally-given glucan. Our published findings showed that simultaneous treatment with glucan and indomethacin caused a very small decrease of phagocytic activity and IL-2 production. Two other tested parameters—blood glucose levels and colon length—were virtually unchanged. In the final but most important part of the study, we found absolutely no mortality, regardless of the type of glucan or the route of glucan administration.

Our conclusion, therefore, was that there are no adverse negative effects of the simultaneous treatment with glucan and a non-steroidal anti-inflammatory drug. This was further verified by the fact that we tested two different routes of glucan administration and four different types of glucan (Vetvicka and Vetvickova, in press). The only possible explanation of the differences between our findings and previously published experiments might be that the original papers used only a type of glucan with particularly peculiar biological properties that cannot be found in any other glucan. However, it might be important to note that both studies (i.e., the positive and negative effects) should be repeated by an additional group of researchers thus allowing one to make a fully sound conclusion.

It is equally important to know that glucan is filed under CFR 21 172.898 as an additive for food and listed by the FDA in the "generally recognized as safe" category with no known contraindications, side effects, or toxicity. In addition, the toxicology of particulate yeast-derived beta glucan WGP™ was studied in detail. The studies were performed under Good Laboratory Practice conditions according to the Organization for Economic Co-operation and Development (OECD) guidelines for the testing of chemicals by an accredited company. In this study, the rats obtained massive daily doses of up to 2 g/kg of glucan. In addition to full biochemical and toxicological assessment, mortality, clinical pathology, functional/behavioral, microscopic and gross observations were performed. No negative effects attributable to the glucan treatment were evident at any dose level. At the same time, the study evaluated the long-term effect of a lower dose (100 mg/kg body weight). Again, no negative effects were observed during the full 13-week study duration. Therefore, the no-observed-adverse effect-level (NOAEL) was determined to be 100 mg/kg body weight/day (Babicek et al., 1997). A similar study also exists for a barley-derived glucan. Delaney and his group evaluated the toxicity of four different doses (from 0.7 up to 7 percent) on a dietary administration for 28 days. The results of the study demonstrated that the consumption of concentrated barley glucan was not associated with any obvious signs of toxicity (Delaney et al., 2003). The dose is what matters the most:

the highest dose this study used corresponded to 336 grams of glucan for a normal-size adult, which is clearly much higher than anybody with a sound mind would ever take. The study rightfully concluded that the consumption of glucan is not likely to cause adverse effects under the conditions of its intended use.

18

GLUCAN IN THE REAL WORLD

This section serves to compare the previous information about glucan to its effects in actual situations. From baldness to cancer, one must keep in mind that glucan is not now, and will probably never be, a cure for everything. However, the internet forum and web pages of individual sellers are filled with comments from happy customers. Some are fabricated, but most are real and, in some cases, glucan's effects can be explained.

Dr. R. R.

About 2 months ago, my wife and I vacationed for a week in Costa Rica. Well, for whatever reason, about a week or so after returning home, I came down with what the doctor diagnosed as "Traveler's Diarrhea", also known as Montezuma's Revenge. The doctor put me on Cipro (an antibiotic), Librax -an intestinal anti-spasmotic, and Lomotil- an antidiarrheal. I also started taking pro-biotics to help restore the natural gut bacteria that the Cipro was destroying. A week later, I'm doing just a very little better, so I go back to the doctor and he prescribes another course of Cipro, plus another anti-infective/anti protozoan called Flagyl. He adds another drug called Bentyl to slow down the intestinal motilitiy.

A week later, I'm still in distress, so I go to see a specialist, - a Gastro-Enterologist. He runs some lab tests to see what's going on. I was tested for Crytosporidium, Giardia, Clostridium Difficile, Ova, and Parasites. Luckily all the tests were negative. He told me to start taking Florastor, another pro-biotic, and see him in 2 weeks for evaluation for a colonoscopy. Well, no real changes and I'm still feeling rotten and want to be done with all this! Then, the miracle! A friend of ours tells us about Beta Glucan. She told us it helps to boost the immune system and she was going to try it for her allergies. I said, "Well I'm going to try it too and see if it will help my intestinal problem!" **Within 2 days** *of starting the Beta Glucan, I was starting to feel better. Less cramping, fewer trips to the bathroom, and just feeling better overall.* **Within 3 days**, *my condition had improved so much that I was able to stop taking the Lomotil and the other meds! Every day I was doing better and better. And the* **only thing different** *in my life was taking the* **Beta Glucan!** *Well, I saw the Gastro-Enterologist yesterday, April 15th, and he said I was discharged from his care for this condition and* **no need for the colonoscopy!** *Again, within 3 days of taking the Beta Glucan the intestinal problem that had been "bugging" me for over 6 weeks was resolving, and within another few days, the condition was essentially gone! So what modern medicine couldn't cure was eliminated by the Beta Glucan.*

Despite all information obtained in the last two decades, our knowledge of glucan action is still not absolutely complete. The experiments testing glucan on diarrhea are lacking, most of all due to the fact that this problems is usually considered too mild (I know, tell it to somebody who is suffering from diarrhea) to deserve the funds for solid testing. However, significant information about glucan acting as both prebiotic and probiotic have appeared lately and I would suspect that glucan help to naturally reach the optimal balance in the colon. Clearly, more studies in this respect have to be done, but this represents another promising area.

PATIENT #1

Mom's MRSA had broken down her immune system to the point that she was developing infection after infection that no antibiotic was helping. I got 2 doses of Beta Glucan according to her weight in her yesterday... the 500 mg... more this morning... and her blood work looks much better already and she said she felt much better too. Jack, they are releasing her today @ 4!!!!!!!!!!!!!!! I never thought it would work this fast. I'm about to cry I'm so happy and grateful to this product!

Glucan has been repeatedly shown to help our immune system fight all types of infection. MRSA is not an exception. In many cases our immune system needs only a small push torward higher activity, and glucan can clearly do just that.

PATIENT #2

I was diagnosed with a rare cancer called schlerosing muco-epidermoid carcinoma last year. The cancer, according to the medical university hospital, was a result of radiation poisoning from the usage of cell phones over the years. I had been taking 1000 mg of beta 1, 3-D Glucan for a year. According to the doctors at the medical university my immune system had, as they characterized it, "surrounded and blocked off the tumor with a wall" preventing the tumor from spreading.

I underwent major surgery removing my parotid gland, the tumor and my facial nerve. I underwent facial reanimation which simply means that the surgeons rebuilt my face with the transfer of tissue and arteries from my arm. The doctors were absolutely amazed at my healing process and commented that I was healing far faster than anyone they had come across over the years. I underwent 33 days of radiation therapy and the doctors were amazed that I had no burning of the skin, due to my use of beta glucan lotion. I also had almost no evidence of scarring from the incision on my neck.

There is no doubt that good glucan strongly supports and activates the immune system. One of the supporting activities is making sure our body can better fight different types of cancer more efficiently. Due to these effects, it also supports healing, particularly from severe types of surgery. Published reports of stimulation of wound healing agree with the effects described by this patient.

Y. S.

I was diagnosed with colorectal cancer in August 2007. Unfortunately, at the time of diagnosis of the 5 cm tumor in my colon, they also found 8 liver metastases on my liver. They operated on the colon to remove the tumor, but I was placed on Folfox chemotherapy for the liver lesions. Luckily for me, the lesions did shrink with chemotherapy and I was able to have 65% of my liver removed in March 2008. The next period we watched and waited to see what was going to happen, then suddenly my tumor marker CEA started to rise again from .9 to 12. A scan in November 2008 revealed 2 more liver lesions so during that month, I was operated again on the liver. This time, however, I was not so lucky as they found peritoneal metastases which is a much poorer prognosis. The surgeon ablated the tumors on my liver and on the peritoneum, but at that stage my chances of a cure were 0. My surgeon suggested I keep positive as miracles do happen, but things really didn't look good.

Whilst all this was going on, the laboratory had sent a sample of my liver to be KRAS tested to see if I would be a suitable candidate for Erbitux. This would be my last chance as the last found of Folfox and then another round of Xeloda had failed to control the disease. I received the news that I do carry the wild type KRAS gene so they started me on Erbitux 15 February 2009. I wasn't able to have che-motherapy with the Erbitux as most patients do, as my oncologist felt I had far too much chemo already. I was a little nervous, as the statistics on Erbitux given on its own is not very promising.

In 2008, I had started the Bill Henderson protocol, but was not able to get some of the vitamins he suggests in his book as they wouldn't deliver to Curacao. I started taking 3 capsules of the Beta Glucan a day on an empty stomach. One hour later I would have breakfast and the rest of my daily vitamins. The tablets are easy to take and to this day I have not had any problems or side effects from them. During the 3rd week of January, I had a CT scan which showed the cancer had now metastasized to both lungs and to many areas mostly in the lower pelvic region as well as the previously treated are near my liver. On 15th February 09 I started Erbitux on its own. I also carried on with the Bill Henderson protocol which included the Beta Glucan. At this stage, I had only been taking them for about 3 weeks.

Last week, after 7 weeks of Erbitux and 10 weeks of taking Beta Glucan and the vitamins in Bill Henderson's protocol, I was scanned again and had the most amazing results. In such a short time, the lung lesions have reduced by 50% and they can only see a very small region of disease on my peritoneal which could even be dead tissue from the previous operation. I am ecstatic with this news as you can imagine. I now have a chance of survival once again, after being thought of as terminal.

I did read an article that mentioned that Beta Glucan can enhance the effectiveness of Erbitux. Well, I am astounded at what has happened. In my wildest dreams I didn't think I would have this result even after a long while of being on treatment. The results are incredible and within such a short time. Thank you so much for sup-plying this fantastic product. I will continue using Beta Glucan for

113

*the rest of my life, even when I am totally CANCER FREE. Thank you
for bringing me this far!*

This is a perfect example of clinical results confirming the idea based on laboratory experiments described in the **Chapter 7. Effects of Glucan on Cancer.** No wonder so many clinical trials are trying to obtain hard data giving a conclusive proof that glucan significantly stimulates the cancerostatic effects of anti-cancer monoclonal antibodies.

L.P.

*Thanks again for sending me the Beta Glucan lotion. I have
used it for one week, and so far I am impressed:*
> 1. *Used the lotion on an abrasion on my hand, next day it
>   had healed up and looked nice and pink instead of
>   being cracked.*
> 2. *Used it on my lower legs, which forever have been very
>   flaky and dry. Now the skin looks nice and smooth.*
> 3. *Used on my face, where I have brown sun spots.
>   It appears the brown spots are starting to fade
>   after just one week's usage of the lotion.*
> 4. *Also used the lotion on my pet, where she was itching,
>   and it seemed to help.*

The effects of glucan on skin are limited so far since researchers have focused more on more serious diseases such as cancer and infections. However, from the limited knowledge of the effects on keratinocytes and from additional data on wound healing, glucan has substantial positive effects. Brown spots are one of the five main signs of aging from the sun and we can only speculate about the role of glucan in the fading of these spots.

J. O., CALIFORNIA

*You have supplied me with the sword & shield that I needed
to battle the giant cancer dragon. It turned out to be much worse
than I thought, if that was even possible. The type of melanoma I got
was the worst kind, and it was more than that. The oncologist head
of the melanoma cancer center said it was the worst he's ever seen in
his professional life. That was why he scheduled all sorts of scans,
PET, x-rays, etc...because I think he thought it has already spread
all over (lungs, brain, liver, etc). I did my background research and*

discovered more recently why. My melanoma registered a 50 mitosis rate. People die from a 0 mitosis rate. The higher the number the worse it is. The highest published on 'net I found was 6. Highest average seems to be around 2 or 3. I'm still not sure of the numbers but after hours of looking online that seems to be it.

So I was battling the biggest melanoma cancer dragon of all the dragons on the planet!! While most others were 2 or 3 feet high, my fire breathing ferocious dragon was 50 feet high. I wanted to give beta glucan and other supplied nutrients sufficient time to work. In fact, I felt like I didn't need surgery. I postponed it twice, before family pressure caused me to go for it. (They all thought I was going to die). After major surgery, the doctor FOUND NOTHING!!!! Thank you thank you!!! Halleluiah!!

Even when your story is not completely usual (and not everybody can expect to be cured from melanoma by glucan only), it provides an interesting view on the effect of a glucan supplement in real life.

L. S. C., COLORADO

I have struggled with fibromyalgia for the past 15 years and have found several supplements that have aided me in enjoying an active lifestyle. When I have a cold or virus, my fibromyalgia flares and the painful symptoms are severe. One of the supplements I take to support my immune system is Beta-1, 3-D glucan. I definitely have fewer colds and flus (many years, not even one). However, if I start having cold or flu symptoms, taking extra Beta glucans reduces not only my cold/flu symptoms, but also prevents my fibromyalgia pain from increasing so dramatically.

There is already so much data demonstrating the effects of glucan on various infections, that the lower incidence (or at least less symptoms) of colds after taking glucan is not surprising. Flu is caused by a virus. With a limited knowledge about glucan and viruses we can only expect to see similar effects. Regarding fibromyalgia, the truth is that this disease is rather complicated. It has been considered by many either a <u>musculoskeletal</u> disease or <u>neuropsychiatric</u> condition. Although there is as yet no cure for fibromyalgia, some treatments have been demonstrated by controlled <u>clinical trials</u> to be effective in reducing symptoms, including medications, behavioral interventions, patient education, and exercise. I would expect that glucan has no direct effect on fibromyalgia, but by reducing the incidence and/or symptoms of cold and/or flu, it reduces the negative effects of these problems on fibromyalgia.

J. C. C.

I have been taking Clariton D for over 15 years and I have tried everything non-medical to help with my allergies nothing worked until now I HAVE NOT TAKEN A CLARITON IN 2 DAYS. Two months ago I found out I had high sugar I went and purchased $289.00 worth of supplements to try to control my sugar without the typical medication. In the mornings is when my sugar seems to be higher than it should, (the norm should be 90 and my sugar has been running at waking up in the morning 158 to 162 for the last two mornings), my sugar has been 112 & 116. This is the only reason I went to look because of what I had read about the product. <u>Something is happening here people.</u>

The effects of glucan on blood sugar are less studied. However, of the studies, some suggested that glucans might reduce blood glucose concentrations after eating by delaying bowel movements so the glucose is absorbed more gradually. Newer studies have shown the direct hypoglycemic activity of natural glucans and synthetic polysaccharides. Some follow-up experiments even suggested that small synthetic oligosaccharides lower the blood sugar similar to natural glucans, which would completely rule out the possible effects via fiber action.

J. G.

I have numerous health issues as many people do when they reach their 50's. I've been excited to start the product, and last Saturday, after we returned home, our mailbox was full of Beta Glucan! Just a couple of days prior I developed an infection at the gum line. The last time that happened was about 8 months ago, and it required a trip to the dentist. He administered laser treatment, and put me on antibiotics and a special mouth rinse. It still took 2 weeks to totally clear up. I took glucan on Saturday and again on Sunday. I was in disbelief on Monday when I woke up, and told my wife that I think something strange was happening! The swelling had gone down 90% and so did the pain. Again, I wanted to be sure, so I hesitated to put this information out there. Well, I obviously took the glucan again on Monday (I start my day with 2 capsules every day!) and yes, much to my own disbelief, the infection was gone; no swelling, no pain, no irritation.

As mentioned repeatedly throughout the book, glucan is well known to support the immune system it its fight against infection. In fact, there is no reference in the scientific literature describing a particular type of infection, where the glucan would be inactive or unhelpful. The gum infection is no exception.

19

GLOSSARY OF TERMS

Angiogenic factors Factors that are critical to the initiation of angiogenesis and maintenance of the vascular network.

Angiogenesis The growth of new blood vessels aiming to feed a tumor.

Antibodies Immune system-related proteins called "immunoglobulins" Each antibody consists of four polypeptides: two heavy chains and two light chains joined to form a "Y" shaped molecule. Antibodies are divided into five major classes, IgM, IgG, IgA, IgD, and IgE, based on their structure and immune function.

Apoptosis Pre-programmed cell death.

Apoptotic cells Cells undergoing the programmed cell death.

Aquaculture Farming of freshwater and saltwater organisms including fish, mollusks and plants. Unlike fishing, aquaculture, also known as "aquafarming" implies the cultivation of aquatic populations under controlled conditions.

BRM

Biological response modifiers.

Carbohydrate

Carbohydrates, or saccharides, are sugars and starches,which provide energy for humans and animals, and cellulose which make up many plant structures. Generally, there are two types of carbohydrates: simple, or *monosaccharides* and complex, or *polysaccharides*.

Carcinogen

A substance that is capable of causing or aggravating cancer in humans or animals. Though a great many things exist that are believed to cause cancer, a substance is only considered carcinogenic if there is significant evidence of its carcinogenicity.

Cellular branch

The immune system consists of two branches : the cellular branch and the humoral one. The cellular branch (also called "cellular immunity" or "cell-mediated immunity") involves the activation of the macrophages, natural killer cells, antigen-specific T-lymphocytes, and the release of various cytokines.

Chemotherapy

A general term for any treatment involving the use of chemical agents (including drugs) to stop cancer cells from growing. Chemotherapy can eliminate cancer cells at sites that are far from the original cancer. As a result, chemotherapy is considered a systemic treatment. Usually, chemotherapy is accompanied with significant negative side effects.

Complement

The thermolabile group of proteins in the blood serum and plasma of humans and animals that (in combination with antibodies) causes the destruction especially of particulate antigens, bacteria and cells. The complement system is a biochemical cascade that helps clear pathogens from an organism. The whole system consists of a number of small proteins found in the blood, normally circulating as inactive precursors. When stimulated, enzymes in the system cleave specific proteins to release different factors and initiate an amplifying cascade of further cleavages. The end-result of this activation cascade is a massive amplification of the response and activation of the cell-killing membrane attack complex. The complement system consists of over 20 proteins and protein fragments.

Corticosterone

Steroid hormone secreted by the outer layer of the adrenal gland. Classed as a glucocorticoid, this hormone helps regulate the conversion of amino acids into carbohydrates and glycogen by the liver, and helps stimulate glycogen formation in the tissues. Corticosterone is similar in structure to the other glucocorticoids *cortisol* and *cortisone*. It is produced in response to the stimulation by the pituitary substance ACTH.

CR3

A receptor of the complement system, a part of the mediated innate immune system. This receptor, referred to as "complement receptor

3", or "CR3", is present on granulocytes, mononuclear phagocytes, and
natural killer (NK) cells and binds
an iC3b fragment of complement.
This CR3 family of receptors is very
important for cell adhesion and cell
mediated cytotoxicity. It is involved
in the binding of glucan molecule.

Cytokine

Small secreted proteins which mediate and regulate immunity, inflammation, and hematopoiesis. They are
produced in response to an immune
stimulus. They generally act over
short distances and short time spans
and at very low concentration. They
act by binding to specific membrane
receptors, which then signal the cell
to alter its behavior. Responses to
cytokines include the increasing or
decreasing expression of membrane
proteins, the proliferation, and secretion of various effector molecules.
"Cytokine" is a general name; other
names include "lymphokine" (cytokines made by lymphocytes), "mo-
nokine" (cytokines made by monocytes), "chemokine" (cytokines with
chemotactic activities), and "interleukin" (cytokines made by one leukocyte and acting on other leukocytes).

Dectin-1

This is the second major receptor for
glucan on the surface of macrophages
and neutrophils. Similarly to CR3, it
mediates the transfer of the signal.

Dendritic cells

These cells are potent antigen-presenting cells possessing the ability to

stimulate naïve T lymphocytes. They comprise a system of leukocytes widely distributed in all tissues. Dendritic cells, often called "pacemakers of the immune reactions", possess a heterogeneous haemopoietic lineage, in that subsets from different tissues have been shown to possess a differential morphology, phenotype and function.

Diabetes

Diabetes is a disease in which the body does not produce or properly use insulin. Insulin is a hormone that is needed to convert sugar, starches and other food into energy needed for daily life. The cause of diabetes remains unclear, although both genetics and environmental factors (such as obesity) appear to play important roles.

Diapedesis

Movement of leukocytes out of the circulatory system, toward the site of tissue damage or infection, via squeezing through the wall of veins and arteries.

DNA

Deoxyribonucleic acid is a nucleic acid that contains the genetic instructions used in the development and functioning of all known organisms.

GALT

Gut-associated lymphoid tissue. Consists of various tissues and organs, such as tonsils, Peyer's patches in the gut or lymphoid aggregates.

Humoral branch

The second part of the immune response. Effector B lymphocytes produce soluble antibodies, which circulate throughout the body and function to eliminate antigens from the organism. This branch of the immune system is known as the "humoral branch" or "humoral immunity". Memory B lymphocytes function to recognize the antigen in future encounters by continuing to express the membrane-bound form of the antibody.

Humic acids

Humic acids represent a group of rather common high molecular weight macromolecules consisting of complex polymeric aromatic structures. They are produced by chemical and microbial degradation of organic matter coming from plants and animals. These compounds can be found in lignite, turf, soil, and drinking water.

Hypercholesterolemia

The presence of excess cholesterol in the blood

Immune system

The immune system is a vast network of cells, tissues, and organs that work together to defend the body against attacks by foreign invaders. It is the immune system's job to keep them out or, failing that, to seek out and kill them. The immune system is amazingly complex. It can recognize and remember millions of different enemies and it can produce secretions (antibodies, cytokines) and cells to match up with and wipe out nearly all of them. If the immune system is crippled, it leaves the body vulnerable to illnesses.

Immunocyte

A cell of the lymphoid lineage which can react with antigens to produce antibodies (B lymphocytes) or to become active in cell-mediated immunity or delayed hypersensitivity reactions; also called "immunologically competent cell" or "immunocompetent cell".

Immunoglobulin

The general term used for antibodies. Depending on their structure, immunoglobulins (Ig) can be divided into IgM, IgG, IgA, IgD and IgE,

Immunomodulation

The change in the body's immune system (either positive or negative), caused by agents that activate (*immunoactivation*) or suppress (*immunosuppression*) its function.

Immunosuppression

Negative adjustment of the immune response.

Indomethacin

A non-steroidal anti-inflammatory drug commonly used to reduce fever, pain, and swelling. It works by inhibiting the production of prostaglandins, which are molecules known to cause these symptoms.

Interleukin

Protein of a group of related proteins made by white blood cells and other cells in the body. Interleukins regulate immune responses.

kDa

A term used for describing the mass of individual molecules. With large molecules, masses are in kilodaltons (kDa), where one kilodalton is 1000 daltons. The precise definition is that it is one twelfth of the mass of an unbound atom of carbon-12 at rest.

Kupffer cells

The resident macrophages of the liver that play an important role in its normal physiology and homeostasis as well as participating in the acute and chronic responses of the liver to toxic compounds. Activation of these cells by toxic agents results in the release of an array of bioactive substances. This activation appears to modulate acute hepatocyte injury as well as chronic liver diseases including hepatic cancer.

Lactosylceramide

Glycoprotein containing a hydrophobic ceramide lipid and hydrophilic sacharidic moieties.

Langerhans cells

Cells of the macrophage lineage, found in the skin. These cells were named after the German physician Paul Langerhans, who first described them. They are involved in various skin infections.

Leukopenia

A low white blood cell count is a decrease in leukocytes circulating in your blood. The benchmark for a low white blood cell count vary slightly among medical practices. In adults it is generally defined as fewer than 3,500 white blood cells per microliter of blood. A low white blood cell count in children varies with age and sex. There are several subtypes of white blood cells, each with different defense activities. If you have a low white blood cell count, you most likely have a decrease in only one type. Leukopenia is often the result of a chemotherapy or radiation therapy.

LPS

Lipopolysaccharide is the major component of the outer membrane of Gram-negative bacteria. The LPS molecule is composed of two biosynthetic entities: the lipid A - core and the O-polysaccharide. Most of the biological effects of LPS are due to the lipid A part. LPS is an exogenous pyrogen (i.e., external fever-inducing compound). LPS also induces strong immune responses, and larger doses can be lethal.

Lymphocyte

A type of immune cell that is made in the bone marrow and is found in the blood and in all the lymph tissues. The two main types of lymphocytes are B lymphocytes and T lymphocytes. B lymphocytes make antibodies, T lymphocytes help control immune responses and are involved in the killing of tumor cells.

Macrophage

Cells of the myeloid lineage. Present in almost every tissue and organ. Their major function is to monitor the circulating streams of body fluid (blood and lymph) and to react adequately to any changes. Their most pronounced function is phagocytosis, which is one of the most important defense mechanisms (i.e, destruction of invaders by ingestion).

Mannan

Polymer of mannose.

Myelopoiesi

The production of myeloid cells in the bone marrow. It might involve the production of all or some cells of the erythroid, megakaryotic or myelocytic lineage.

Mitogen	An agent that stimulates cell division and lymphocyte transformation.
mRNA	*Messenger ribonucleic acid* is a molecule of RNA encoding a chemical blueprint for individual proteins.
Neutrophil	A type of white blood cells or leukocytes which form an early line of defense against bacterial infections. Neutrophils represent the largest part of the leukocytes, and are produced in high numbers in response to infection, trauma, stress or other stimuli. They circulate around the blood, waiting to be called to a site where damage is happening. Once there, they kill the invading bacteria and other noxious substances. The method they use to kill invaders starts with phagocytosis and is followed by the release of various highly toxic substances such as hydrolytic enzymes of peroxidases. Neutrophils are very short lived, lasting anything from a few hours to a few days.
NK cell	Natural killer (NK) cells represent a small population of the T lymphocytes. They examine other cells for the presence of certain molecules that identify the cell as a cancerous cell or a virus-infected cell. If such a cell is found, the NK cell deploys a tentacle with venom sacks and attaches to the suspect cell. An injection of venom follows, resulting in the death of the target cell.
NOAEL	*No observed adverse effect level.* It denotes the level of exposure,

determined by experiment, at which there is no biologically or statistically significant increase (e.g. alteration of morphology, functional capacity, development or life span) in the frequency or severity of any adverse effects in the exposed population when compared to its appropriate controls.

Non-Hodgkin's lymphoma Also known as "NHL" or sometimes just "lymphoma", is a cancer that starts in the cells of the lymph system, which is part of the body's immune system. There are two main types of lymphomas– "Hodgkin lymphoma" (also known as "Hodgkin's disease") and "non-Hodgkin lymphomas".

Parenteral When a substance is given by routes other than via the digestive tract (intro-peritoneally, sub-cutaneously, etc.).

Peyer's patch The group of Peyer's glands on the inner wall of the small intestine. It usually has an elongated shape and its function is related to the immune system. It plays an important role in the transfer of biologically active molecules through the intestinal wall.

Phagocytosis Originally the means by which larger materials move into the cells. Probably one of the most widely occurring cellular functions. The process originally served in the normal feeding cycle of unicellular animals, later it acquired a new reason. The term came from Greek *"phagein"* meaning "to eat" and is used to describe the intake of solid particles such as red blood cells or bacteria. This process represents an extremely important part of cellular defense mechanisms.

Placebo

A pharmacologically inert substance (such as starch or glucose tablet) that produces an effect similar to what would be expected of a pharmacologically active substance. The placebo effect is the measurable, observable, or felt improvement in health not attributable to a medication or treatment that has been administered.

Pluripotent cell

Primordial cells that may still differentiate into various specialized types of tissue elements.

Polysaccharide

Complex saccharides. Polymers made up of many monosaccharides (simple saccharides) joined together by glycosidic bonds.

Prophylactic treatment

Treatment designed and used to prevent a particular disease from occurring.

Receptor

Protein molecule, embedded in either the cytoplasm or plasma membrane of a cell, to which a signal molecule will bind. A molecule, which binds to a receptor is called a ligand. When such binding occurs, the receptor undergoes a conformational change usually resulting in a cellular response.

Sepsis

A condition in which the body is fighting a severe infection that has spread via the bloodstream. Sepsis is a serious infection usually caused by bacteria — which can originate in many body parts, such as the lungs, intestines, urinary tract, or skin — that make toxins causing the immune system to attack the body's own organs and tissues.

Thimerasol

A mercury-containing organic compound. Since the 1930s, it has been widely used as a preservative in a number of products, including many vaccines, to help prevent potentially life threatening contamination with harmful microbes. Because of potential toxicity, Thimerosal has been removed from or reduced to trace amounts in all vaccines routinely recommended for children of 6 years of age and younger, with the exception of inactivated influenza vaccine.

Theophylline

Also known as dimethylxanthine, is a methylxanthine drug used in therapy for respiratory diseases such as COPD and asthma under a variety of brand names

Virus

Genetic entities that lie somewhere in the grey area between living and non-living states. Viruses depend on the host cells that they infect to reproduce. When found outside of host cells, viruses exist as a protein coat, sometimes enclosed within a membrane. The coat encloses either DNA or RNA which codes for the virus elements. While in this form outside the cell, the virus is metabolically inert.

Zymosan

An insoluble mannose-rich cell wall

polysaccharide of yeast. Besides being a potent activator of macrophages, zymosan also induces the release of cytokines from neutrophils and pro-inflammatory cytokines in immune cells.

20

SCIENTIFIC REFERENCES

The scientific papers used in this book represent only a tiny portion of the huge amount of published reports on glucans and by no means represent a full or even comprehensive list. Readers eager to read scientific reports should start with some of these excellent reviews (Kogan, 2000, Schepetkin and Quinn, 2006, Novak and Vetvicka, 2008, Vetvicka and Novak, 2011).

AbuMweis S. S., Jew, S., Ames, N. P.: ß—Glucan from barley and its lipid-lowering capacity: a meta-analysis of randomized, controlled trials. *Eur. J. Clin. Nutrition*, 64: 1472-1480, 2010.

Akramiene, D., Aleksandraviciene, C., Grazeliene, G., Zalinkevicius, R., Suziedelis, K., Didziapetriene, J., Simonsen, U., Stankevicius, E., Kevelaitis, E.: Potentiating effects of β-glucans on photodynamic therapy of implanted cancer cells in mice. Tohoku *J. Exp. Med.*, 220: 299-306, 2010.

Alwis, K. U., Mandryk, J., Hocking, A. D.: Exposure to biohazards in wood dust: bacteria, fungi, endotoxins, and (1-->3)-beta-D-glucans. *Appl. Occupat. Environment. Hyg.*, 14: 598-608, 1999.

Aziz, A., Poinssot, B., Daire, X., Adrian, M., Bezier, A., Lambert, B., Joubert, J. M., Pugin, A.: Laminarin elicits defense responses in grapevine and induces protection against *Botrytis cinerea* and *Plasmopara viticola*. *Molecular Plant-Microbe Interactions,* 16: 1118-1128, 2003.

Babicek, K., Cechova, I., Simon, R. R., Hardwood, M., Cox, D. J.: Toxicological assesment of a particulate yeast (1,3/1,6)-β-D-glucan in rats. *Food Chem. Toxicol.,* 45: 1719-1730, 1997.

Babincova, M., Bacova, Z., Machova, E., Kogan, G.: Antioxidant properties of carboxymethylglucan: comparative analysis. *J. Med. Food,* 5: 79-83, 2002.

Babineau,T. J., Marcello, P., Swalis, W., Kenler, A., Bistrian, B., Forse, R. A. Randomized phase I/II trial of a macrophage-specific immunomodulatory (PGG-glucan) in high-risk surgical patients. *Annals of Surgery,* 220: 601-609,1994.

Badulescu, M.-M., Apetrei, N. S., Lupu, A.-R., Cremer, L., Szegli, G., Moscovici, M., Mocanu, G., Mihai, D., Calugaru, A.: Curdlan derivatives able to enhance cytostatic drugs activity on tumor cells. *Roum. Arch. Microbiol. Immunol.,* 68: 201-206, 2009.

Bertelli, A. A. E., Ferrara, F., Diana, G., Fulgenzi, A., Corsi, M., Ponti, W., Ferrero, M. E., Bertelli, A.: Resveratrol, a natural stilbene in grapes and wine, enhances intraphagocytosis in human promonocytes: a co-factor in antiinflammatory and anticancer chemopreventive activity. *Int. J. Tissue React.,* 21:93-104, 1999.

Bhat, K. P. L., Pezzutom J. M.: Cancer chemopreventive activity of resveratrol. *Ann. N.Y. Acad. Sci.,* 957:210-229, 2002.

Bohn, J. A. and J. N. BeMiller. (1-3)-β-D-glucans as biological response modifiers: a review of structure-functional activity relatioships. *Carbohydrate Polymers,* 28: 3, 1995.

Bowers, G. J., Patchen, M. L., MacVittie, T. J., Hirsch, E. F., Fink, M. P.: Glucan enhances survival in an intraabdominal infection model. *J. Surg. Res.*, 47: 183-188, 1989.

Brasnyo, P., Molnar, G. A., Mohas, M., Marko, L., Laczy, B., Cseh, J., Mikolas, E., Andras Szijarto, I., Merei, A., Halmai, R., Meszaros, L. G., Sumegi, B., Wittmann, I.: Resveratrol improves insulin sensitivity, reduces oxidative stress and activates the Akt pathway in type 2 diabetic patients. *Br. J. Nutr.*, 2011, in press.

Browder, W., Williams, D., Pretus, H. A., Enrichsen, F., Mao, P., Franchello, A.: Beneficial effect of enhanced macrophage function in the trauma patients. *Ann. Surg.*, 211: 605-613, 1990.

Breivik, T., Opstad, P. K., Engstad, R., Gundersen, G., Gjermo, P., Preus, H. L.: Soluble β-1,3/1,6-glucan from yeast inhibits experimental periodontal disease in Wistar rats. *J. Clin, Periodont.*, 32: 347-362, 2005.

Chen, K. L., Weng, B. C., Chang, M. T., Liao, Y. H., Chen, T. T., Chu, C.: Direct enhancement of the phagocytic and bactericidal capability of abdominal macrophage of chicks by β-1,3-1,6-glucan. *Poultry Sci.*, 87: 2242-2249, 2008.

Chihara, G., Maeda, Y. Y., Hamuro, J., Sasaki, T., Fukuoka, F.: Inhibition of mouse sarcoma 180 by polysaccharides from *Lentinus edodes* (Berk.) sing. *Nature,* 222: 687-688, 1969.

Csiszar, A., Smith, K., Labinsky, N., Orosz, Z., Rivera, A., Ungvari, Z.: Resveratrol attenuates TNF-α–induced activation of coronary arterial endothelial cells: role of NF-κB inhibition. *Am. J. Physiol. Heart Circ. Physiol.*, 291: H1694-H1699, 2006.

Dalmo, R. A., Ingebrigtsen, K., Bogwald, J.: Non-specific defence mechanisms in fish, with particular reference to the reticuloendothelial system (RES). *J. Fish Dis.*, 20: 241–273, 1997.

Descroix, K., Vetvicka, V., Laurent, I., Jamois, F., Yvin, J.-C.: New oligo-ß-glucan derivatives as immunostimulating agents. *Bioorganic Med.* Chem., 18: 348-357, 2010.

Delatte, S. J., Evans, J., Hebra, A., Adamson, W., Othersen, H. B., Tagge, E. P.: Effectiveness of beta-glucan collagen for treatment of partial-thickness burns in children. *J. Pediat. Surgery,* 36: 112-118, 2001.

Delaney, B., Carlson, T., Frazer, S., Zheng, T., Hess, R., Ostergren, K., Kierzek, K., Haworth, J., Knutosn, N., Junker, K., Jonker, D.: Evaluation of the toxicity of concentrated barley β-glucan in a 28-day feeding study in Wistar rats. *Food Chem. Toxicol.,* 41: 477-487, 2003.

DeLuca A. J., Brogden K. A., French A. D.: Agglutination of lung surfactans with glucan. *Br. J. Ind. Med.,* 49: 755-760, 1992.

Demir, G., Klein, H. O., Mandel-Molinas, N., Tuzuner, N.: Beta glucan induces proliferation and activation of monocytes in peripheral blood of patients with advanced breast cancer. *Internal. Immunopharm.,* 7: 112-116, 2007.

Di Luzio, N. R., Riggi, S. J.: The effects of laminarin, sulfated glucan and oligosaccharides of glucan on reticuloendothelial activity. *J. Reticuloendothel. Soc.,* 8: 465-473, 1970.

Ditteova, G., Velebny, S., Hrckova, G.: Modulation of liver fibrosis and pathological changes in mice infected with Mesocestoides corti (M. vogae) after administration of glucan and liposomized glucan in combination with vitamin *C. J. Helmintol.,* 77: 219-226, 2003.

Do Amaral, C. L., Francescato, H. D., Coimbra, T. M., Costa, R. S., Darin, J. D., Antunes, L. M., Bianchi, M. D.: Resveratrol atteanuates cisplatin-induced nephrotoxicity in rats. Arch. Toxicol., 82:363-370, 2007.

Ebina,T., Fujimiya,Y.: Antitumor effect of a peptide-glucan preparation extracted from *Agaricus blazei* in a double-grafted tumor system in mice. *Biotherapy,* 11: 259-265, 1998.

Eicher, S. D., McKee, C. A., Carroll, J. A., Pajor, E. A.: Supplemental vitamin C and yeast cell wall β-glucan as growth enhancers in newborn pigs and as immunomodulators after an endotoxin challenge after weaning. *J. Anim. Sci.,* 84: 2352-2360, 2006.

Falchetti, R., Fuggetta, M.P., Lanzilli, G., Tricarico, M., Ravagnan, G.: Effects pf resveratrol on human immune cell function. *Life Sciences* 70:81-96, 2001.

Fogelmark, B., Sjöstrand, M., Rylander, R.: Pulmonary inflammation-induced by repeated inhalations of beta(1,3)-D-glucan and endotoxin. *Int. J. Exp. Pathol.,* 75: 85-90, 1994.

Fornusek, L., Vetvicka, V.: Immune System Accessory Cells. *CRC Press,* 1992.

Gibson, G. R., Roberfroid, M. B.: Dietary modulation of the human colonic microbiota: introducing the concept of prebiotics. *J. Nutr.,* 125: 1401-1412.

Gordon, M., Guralnik, M., Kaneko, Y., Mimura, T., Goodgame, J., DeMarzo, C., Pierce, D., Baker, M., Lang, W.: A phase II controlled study of a combination of the immune modulator, lentinan, with didanosine (ddI) in HIV patients with CD4 cells of @))-500 mm^{-3}. *J. Med.,* 26: 193-207, 1995.

Gulcelik, M. A., Dincer, H., Sahin, D., Demir, O. F., Yenidogan, E., Alagol, H.: Glucan imrpves impaired wound healing in diabetic rats. Wounds – *A Compendium Clin. Res. Pract.,* 22: 12-16, 2010.

Guo, Z., Hu, Y., Wang, D., Ma, Z., Zhao, X., Zhao, B., Wang, J., Lie, P.: Sulfated modification can enhance the adjuvanticity of lentinan and improve the immune effects of ND vaccine. *Vaccine 27*: 660-665, 2009.

Gupta, S., Kannappan, R., Reuter, S., Kim, J. H., Aggarwal, B. B.: Chemosensitization of tumors by resveratrol. *Ann. N.Y. Acad. Sci.*, 1215: 150-160, 2011.

Haladova, E., Mojzisova, J., Smrco, P., Ondrejkova, A., Vojtek, B., Prokes, M., Petrovova, E.: Immunomodulatory effect of glucan on specific and nonspecific immunity after vaccination in puppies. *Acta Vet. Hung.*, 59: 77-86, 2011.

Hamano, K., Gohra, H., Katoh, T., Fujimora, Y., Zempo, N., Esato, K.: The preoperative administration of lentinan ameliorated the impairment of natural killer activity after cardiopulmonary bypass. *Int. J. Immunopharmac.*, 21: 531-540, 1999.

Harada T, Ohno N.: Contribution of dectin-1 and granulocyte macrophage-colony stimulating factor (GM-CSF) to immunomodulating action of β-glucan. *Int. Immunopathol.*, 8: 556-566, 2008.

Harnack, U., Eckert, K., Fichtner, I., Pecher, G.: Oral administration of a soluble 1-3, 1-6 β-glucan during prophylactic survivin peptide vaccination diminishes growth of a B cell lymphoma in mice. *Int. Immunopharmacol.*, 9: 1298-1303, 2009.

Hong, F., Hansen, R. D., Yan, J., Allendorf, D. J., Baran, J. T., Ostroff, G. R., Ross, G. D.: β-Glucan functions as an adjuvant for monoclonal antibody immunotherapy by recruiting tumoricidal granulocytes as killer cells. *Cancer Res.*, 63: 9023-9031, 2003.

Hong, F., Yan, J., Baran, J. T., Allendorf, D. J., Hansen, R. D., Ostroff, G. R., Xing, P. X., Cheung, N. K., Ross, G. D.: Mechanism by which orally administered beta-glucans enhance the tumoricidal activity of antitumor monoclonal antibodies in murine tumor models. *J. Immunol.*, 173: 797-806, 2004.

Horvatova, E., Eckl, P. M., Bresgen, N., Slamenova, D.: Evaluation of genotoxic and cytotoxic effects of H_2O_2 and DMNQ on freshly isolated rat hepatocytes; protective effects of carboxymethyl chitin-glucan. *Neuroendocrinol. Lett.*, 29: 644-648, 2008.

Hotta, H., Hagiwara, K., Tabata, K., Ito, W., Homma, M.: Augmentation of protective immune responses against Sendai virus infection by fungal polysaccharide schizophyllan. *Int.J.Immunopharmacol.*, 15: 55-60, 1993.

Itoh, W., Sugawara, I., Kimura, S., Tabata, K., Hirata, A., Kojima, T., Mori, S., Shimada, K.: Immunopharmacological study of sulfated schizophyllan (SPG). I. Its action as a mitogen and anti-HIV agent. *Int. J. Immunopharmacol.*, 12: 225-233, 1990.

Jang, M., Cai, L., Udeani, G. O., Slowing, K. V., Thomas, C. F., Beecher, C. W. W., Fong, H. H. S., Farndworth, N. R., Kinghorn, A. D., Mehta, R. G., Moon, R. C., Pezzuto, J. M.: Cancer chemopreventive activity of resveratrol, a natural product derived from grapes. *Science 275*: 218-220, 1997.

Kabat, E.A. Structural Concepts in Immunology and Immunochemistry. *Holt, Rinehart and Winston, New York,* 1976.

Kaibara, N., Soejima, K., Nakamura, T., Inokuchi, K.: Postoperative long term chemotherapy for advanced gastric cancer. *Jpn. J. Surg.*, 6: 54-59, 1976.

Kernodle, D. S., Gates, H., Kaiser, A. B.: Prophylactic anti-infective activity of poly-[1-6]-β-D-glucopyranosyl-[1-3]-β-D-glucopyranose glucan in guinea pig model of staphylococcal would infection. *Antimicrobial Agents Chemotherap.*, 42: 545-549, 1998.

Kida, K., Inoue, T., Kaino, Y., Goto, Y., Ikeuchi, M., Ito, T., Matsuda, H., Elliott, R. B.: An immunopotentiator of β-1,6;1,3 D-glucan prevents diabetes and insulitis in BB rats. *Diabetes Res. Clin. Prctice*, 17: 75-79, 1992.

Kim, H. D., Cho, H. R., Moon, S. B., Shin, H. D., Yang, K. J., Park, B. R., Jang, H. J., Kim, L. S., Lee, H. S., Ku, S. K.: Effects of β-glucan from *Aureobasidium pullulans* on acute inflammation in mice, *Arch. Pharm. Res.*, 40: 323-328, 2007.

Kirmaz, C., Bayrak, P., Yilmaz, O., Yuksel, H.: Effect of glucan treatment on the Th1/Th2 balance in patients with allergic rhinitis: a double-blind placebo-controlled study. *Eur. Cytokine Netw.*, 16: 128-134, 2005.

Klarzynski, O., Plesse, B., Joubert, J. M., Yvin, J.-C., Kopp, M., Kloareg, B., Fritig, B.: Linear β-1,3 glucans are elicitor of defense responses in tobacco. *Plant Physiol.*, 124: 1027-1037, 2000.

Kogan, G. (1-3,1-6)-β-D-glucans of yeast and fungi and their biological activity. In: *Studies in Natural Products Chemistry, Vol. 23.*, Ed. Atta-ur-Rahman, Elsevier, Amsterdam, p. 107., 2000.

Kouigias, P., Wei, D., Rice, P. J., Ensley, H. E., Kalbfleisch, J., Williams, D. L, Browder, I. W.: Normal human fibroblasts express pattern recognition receptors for fungal (1-3)-β-D-glucans. *Inf.Immun.*, 69: 3933-3938, 2001.

Krakowski, L., Krzyzanowski, J., Wrona, Z., Siwicku, A.K.: The effect of nonspecific immunostimulation of pregnant mares with 1,3/1,6 glucan and levamisole on the immunoglobulins levels in colostrum, selected indices of nonspecific cellular and humoral immunity in foals in neonatal and postnatal period. *Vet. Immunol. Immunopathol.*, 68: 1-11, 1999.

Kraft, T. E., Parisotto, D., Schempp, C., Efferth, T.: Fighting cancer with red wine? Molecular mechanisms of resveratrol. *Crit. Rev. Food Sci. Nutrition*, 49: 782-799, 2009.

Kubo, K., Nanba, H.: The effect of maitake mushrooms on liver and serum lipids. *Alternative Therapies*, 5: 62-66, 1996.

Kukan, M., Szatmary, Z., Lutterova, M., Kuba, D., Vajdova, K., Horecky, J.: Effects of sizofiran on endotoxin-enhanced cold ischemia-reperfusion injury of the rat liver. *Physiol. Res.*, 53: 431-437, 2004.

Kurashige, S., Akuzawa, Y., Endo, F.: Effects of Lentinus edodes, Grifola frondosa and Pleurotus ostreatus administration on cancer outbreak, and activities of macrophages and lymphocytes in mice treated with a carcinogen, N-butyl-N-butanolnitrosoamine. *Immunopharmacol. Immunotoxicol.*, 19: 175-183, 1997.

Li, Ch., Ha, T., Kelley, J., Gao, X., Qiu, Y., Kao, R. L., Browder, W., Williams, D. L.: Modulating Toll-like receptor mediated signaling by (1-3)- β-D-glucan rapidly induces cardioprotection. *Cardiovasc. Res.*, 61: 538-547, 2004.

Magnani, M., Castro-Gomez, R. H., Aoki, M. N., Gregorio, E. P., Libos, F., Watanabe, M. A. E.: Effects of carboxymethyl-glucan from *Saccharomyces cerevisiae* on the peripheral blood cells of patients with advanced prostate cancer. *Exp. Therap. Med.*, 1: 859-862, 2010.

Mai, T. T. T., Igarashi, K., Hirunuma, R., Takasaki, S., Yasue, M., Enomoto, S., Kimura, S., van Chuyen, N.: Iron absorption in rats increased by yeast glucan. *Biosci. Biotechnol. Biochem.*, 66: 1744-1747, 2002.

Mantovani, M. S., Bellini, M. F., Angeli, J. P. F., Oliveira, R. J., Silva, A. F., Ribeiro, L. R.: β-Glucans in promoting health: prevention against mutation and cancer. *Mut. Res.*, 2007.

Mikherjee, S., Dudley, J. I., Das, D. K.: Dose-dependency of resveratrol in providing health benefits. *Dose Response*, 8:478-500, 2010.

Mitsou, E. K., Panopoulou, N., Turunen, K., Spollotis, V., Kyriacou, A.: Prebiotic potential of barley derived β-glucan at low intake levels: A randomised, double-blinded, placebo-controlled clinical study. *Food Res. Int.*, 43: 1086-1092, 2010.

Modak, S., Koehne, G., Vickers, A., O'Reilly, M. J., Cheung, N-K. V.: Rituximab therapy of lymphoma is enhanced by orally administered (1-3), (1-4)-D- β-glucan. *Leukemia Res.*, 29: 679-683, 2005.

Mohagheghpour, N., Dawson, M., Hobbs, P., Judd, A., Winant, R., Dousman, L., Waldeck, N., Hokama, L., Tuse, D., Kos, F., Benike, C., Engleman, E.: Glucans as immunological adjuvans. In: *Immunobiology of Proteins and Peptides VIII*, Eds. Attasi, M. A., Bixler, G. S., Plenum Press, New York, pp. 13-22, 1995.

Murphy, E. A., Davis, J. M., Brown, A. S., Carmichael, M. D., Ghaffar, A., Mayer, E. P.: Oat β-glucan effects on neutrophil respiratory burst activity following exercise. *Medicine Sci, Sports Exercise, 39*: 639-644, 2007.

Murphy, E. A., Davis, J. M., Brown, A. S., Carmichael, M. D., Carson, J. A., Van R. N., Fgaffar, A., Mayer, E. P.: Benefit of oat beta-glucan on resporatory infection following exercize stress. *Am. J. Physiol. Regul. Integr. Comp. Physiol.*, 294: R1593-R1599, 2008.

Nakao, I., Uchino, H., Kaido, I., Kimuira, T., Goto, T., Kondo, T., Takino, T., Taguchi, T., Nakajima, T., Fujimoto, S., Miyazaki, T., Miyoshi, A., Yachi, A., Yoshida, K., Ogawa, N., Furue, H.: Clinical evaluation of schizophyllan (SGP) in advanced gastric cancer – a randomized comparative study by an envelop method. *Jpn. J. Cancer Chemotherap.*, 10: 1146-1159, 1983.

Nanba, H.: Activity of Maitake D-fraction to inhibit carcinogenesis and metastasis, *Ann.NY Acad.Sci.*, 768: 243-244, 1995.

Nameda, S., Miura, N. N., Adachi, Y., Ohno, N.: Antibiotics protect against septic shock in mice administered β -glucan and indomethacin. *Microbiol. Immunol.*, 51, 851-859, 2007 a.

Nameda, S., Miura, N. N., Adachi, Y., Ohno, N.: Lincomycin protects mice from septic shock in β-glucan-indomethacin model. *Biol. Pharm. Bull.*, 30: 2312-2316, 2007 b.

Ngamkala, S., Futami, K., Endo, M., Maita, M., Katagiri, T.: Immunological effects of glucan and Lactobacillus rhamnosus GG, a probiotic bacterium, on Nile tilapiua *Orechromis niloticus* intestine with oral *Aeromonas* challenges. *Fish Sci.*, 76: 833-840, 2010.

Niculescu, F. H., Rus, G., Retegan, M., Vlaicu, R.: Persistent complement activation on tumor cells in breast cancer. *Am. J. Pathol.*, 140: 1039-1043, 1992.

Nio, Y., Tshuchitani, T., Imai, S., Shiraishi, T., Kan, N., Ohgaki, K., Tobe, T.: Immunomodulating effects of oral administration of PSK. II. Its effects on mice with cecal tumors. *Nippon Gan Chiryo Gakkai Shi*, 23: 1068-1087, 1988.

Novak, M., Vetvicka, V.: Beta-glucans, history, and the present: Immunomodulatory aspects and mechanisms of action. *J. Immunotoxicol.*, 5: 47-57, 2008.

Nosalova, V., Bobek, P., Cerna, S., Galbavy, S., Stvrtina, S.: Effects of Pleuran (β-glucan isolated from *Pleurotus ostreatus*) on experimental colitis in rats. *Physiol. Res.*, 50: 575-581, 2001.

Ohno, N.: Detrimental effects of β-glucan. In: Vetvicka, V., Novak, M. (Eds.), *Biology and Chemistry of Beta Glucan - Volume 1: Beta Glucans - Mechanisms of Action,* Bentham Science Publishers, 2011, pp. 68-81.

Ohno, N., Kurachi, Y., and Yadomae, T.: Physicochemical properties and antitumor activities of carboxymethylated derivatives of glucan from *Sclerotinia sclerotiorum. Chem. Pharmacol. Bull.,* 36: 1198-1125, 1988.

Ohno, N., Miura, N.N., Nakajima, M., Yadomae, T.: Antitumor 1,3-beta-glucan from cultured fruit body of *Sparassis crispa. Biol. Pharm. Bull.,* 23: 866-872, 2000.

Patchen, M. L., MacVittie, T. J., Brook, I.: Glucan-induced hemopoietic and immune stimulation: therapeutic effects in sublethally and lethally irradiated mice. *Methods Find. Exp. Clin. Pharmacol.,* 8: 151-155, 1986.

Pillai, R., Redmond, M., Roding, J.: Anti-wrinkle therapy: significant new findings in the non-invasive cosmetic treatment of skin wrinkles with beta-glucan. *IFSCC Magazine,* 8: 1-6, 2005.

Portera, C. A., Love, E. J.: Effect of macrophage stimulation on collagen biosynthesis in the healing wound. *Am. Surgeon,* 63: 125-131, 1997.

Reyna-Villasmil, N., Bermudez-Pirela, V., Mengual-Moreno, E., Arias, N., Cano-Ponce, C., Leal-Gonzales, E., Souki, A., Inglett, G. E., Israili, Z.H., Hernandez-Hernandez, R., Valasco, M., Arraiz, N.: Oat-derived β-glucan significantly improves HDLC and diminished LDLC and non-HDL cholesterol in overweight individuals with mild hypercholesterolemia. *Am. J. Therapeut.,* 14: 203-212, 2007.

Rice, P. J., Adams, E. L., Ozment-Skelton, T., Gonzalez, A. J., Goldman, M. P., Lockhart, B. E., Barker, L. A., Breuel, K.

F., DePonti, W. K., Kalbfleisch, J. H., Ensley, H. E., Brown, G. D., Gordon, S., Williams, D. L.: Oral delivery and gastrointestinal absorption of soluble glucans stimulate increased resistance to infectious challenge. *Pharmacol. Exp. Ther.,* 314: 1079-1086, 2005.

Riles, W. L., Erickson, J., Nayyar, S., Atten, M. J., Attar, B. M., Hollan, O.: Resveratrol engages selective apoptotic signals in gastric adenocarcinoma cells. *W.J. Gastroenterol.,* 21:5628-5634, 2006.

Rosburg, V., Boylston, AT., White, P.: Viability of bifidobacteria strains in yogurt with added oat beta-glucan and corn starch during cold storage. *J. Food Sci.,* 75: C439-C444, 2010.

Ross, G. D., Vetvicka, V.: CR3 (CD11b/CD18): A phagocyte and NK cell membrane receptor with multiple ligand specificities and functions. *Clin. Exp. Immunol.,* 92: 181-184, 1993.

Ross, G. D., Vetvicka, V., Yan, J., Xia, Y., Vetvickova, J.: Therapeutic intervention with complement and β-glucan in cancer. *Immunopharmacology,* 42: 61-74, 1999.

Rovensky, J., Stancikova, M., Svik, K., Bauerova, K., Jurcovicova, J.: The effects of beta-glucan isolated from Pleurotus ostrateus on methotrexate treatment in rats with adjuvant arthritis. *Rheumatol. Int.,* 31: 507-511, 2011.

Sakurai, T., Ohno, N., Yadomae, T.: Intravenously administered (1->3)-β-D-glucan, SSG, obtained from *Sclerotinia sclerotiorum* IFO 9395 augments murine peritoneal macrophage functions *in vivo. Chem. Pharm. Bull.(Tokyo),* 40: 2120-2124, 1992.

Sanchez, D., Quinones, M., Moulay, O., Muguerza, B., Miguel, M., Aleixandre, A.: Soluble fiber-enriched diets improve inflammation and oxidative stress biomarkers in Zucker fatty rats. *Pharmacol. Res.*, in press.

Sato, M., Sano, H.: Direct binding to Toll-like receptor 2 to zymosan, and zymosan-induced NF-kappa B activation and TNF-alpha secretion are down-regulated by lung collection surfactant protein. *J. Immunol.*, 171, 417-425, 2003.

Schepetkin, I. A., Quinn, M. T.: Botanical polysaccharides: macrophage immunomodulation and therapeutical potential. *Int.Immunopharmacol.*, 6: 317-333, 2006.

Sener, G., Toklu, H., Ercan, F., Erkanh, G.: Protective effect of β-glucan against oxidative organ injury in a rat model of sepsis. *Int. Immunopharmacol.*, 5: 1387-1396, 2005.

Shear, M. J., Turner, F. C., Perrault, A., Shovelton, T.: Clinical treatment of tumors. V. Isolation of the hemorrhage-producing fraction from *Serratia marcescens (Bacillus prodigiosus)* culture filtrates. *J. Natl. Cancer Inst.*, 4: 81-97, 1943.

Soto, E., Ostroff, G.: Use of –glucans for drug delivery application. Detrimental effects of β-glucan. In: Vetvicka, V., Novak, M. (Eds.), *Biology and Chemistry of Beta Glucan - Volume 1: Beta Glucans - Mechanisms of Action*, Bentham Science Publishers, 2011, pp. 48-67.

Spruit, N. E., Visser, A. T., Leenen, L. P.: A systematic review of randomized controlled trials exploring the effects of immunomodulative interventions on infection, organ failure, and mortality in trauma patients. *Crit. Care*, 14: R150, 2010.

Stashenko, P., Wang, C. Y., Riley, E., Wu, Y., Osgtroff, G., Niederman, R.: Reduction of iinfection-stimulated periapical bone resorption by the biological response modifier PGG glucan. *J. Dental Res.*, 74: 323-330, 1995.

Stiwari, U., Cummins, E.: Mata-analysis of the effects of beta-glucan intake on blood cholesterol and glucose levels. *Nutrition*, in press.

Stratford, M. Another brick in the wall? Recent developments concerning the yeast cell envelope. *Yeast*, 10:1741-1752, 1994.

Sugiyama, A., Hata, S., Suzuki, K., Yoshida, E., Nakano, R., Mitra, S., Arashida, R., Asayama, Y., Yabuta, Y., Takeuchi, T.: Oral administration of paramylon, a beta-1,3,-D-glucan isolated from *Euglena gracilis Z* inhibits development of atopic dermatitis-like skin lesions in NC/Nga mice. *J. Vet. Med. Sci.*, 72: 755-763, 2010.

Suzuki, I., Sakurai, T., Hashimoto, K., Oikawa, S., Masuda, A., Ohsawa, M., Yadomae, T.: Inhibition of experimental pulmonary metastasis of Lewis lung carcinoma by orally administered β-glucan in mice. *Chem.Pharm.Bull. (Tokyo)*, 39: 1606-1608, 1991.

Synytsya, A., Mickova, K., Synytsya, A., Jablonsky, I., Spevacek, J., Erban, V., Kovarikova, E., Copikova, J.: Glucans from fruit bodies of cultivated mushrooms *Pleurotus ostreatus* and *Pleurotus eryngii*: Structure and potential prebiotic activity. *Carbohydrate Pol.*, 76: 548-556, 2009.

Szymanska-Czerwinska, M., Bednarek, D.: Effect of tylosin and prebiotics on the selected humoral immunological parameters in calves. *Med. Weteryn.*, 67: 275-278, 2011.

Takahasmi H, Ohno N, Adachi Y, Yadomae T. Association of immunological disorders in lethal side effects of NSAIDs on β-glucan-administered mice. *FEMS Immunol. Med. Microbiol.*, *31*, 1-14, 2001.

Takeshita, K., Saito, N., Sato, Y., Maruyama, M., Sunagawa, M., Habu, H., Endo, M.: Diversity of complement activation by lentinan, an antitumor polysaccharide, in gastric cancer patients. *Nippon Geka Gakkai Zasshi*, 92: 5-11, 1991.

Thornton, B. P., Vetvicka, V., Pitman, M., Goldman, R. C., Ross, G. D.: Analysis of the sugar specificity and molecular location of the beta-glucan-binding lectin site of complement receptor type 3 (CD11b/CD18). *J. Immunol.,* 156: 1235-1246, 1996.

Tietyen J. L, Nevins D. J, Schneeman B. O.: Characterization of the hypocholesterolemic potential of oat brand. *FASEB,* 4: A527, 1990.

Tsoni, S. V., Brown, G. D.: β-Glucans and Dectin-1. *Ann. N.Y. Acad. Sci.,* 1143: 45-60, 2008.

Turnbull, J. L., Patchen, M. L., Scadden, D. T.: The polysaccharide, PGG-glucan, enhances human myelopoiesis by direct action independent of and additive to early-acting cytokines. *Acta Haematol.,* 103: 66-71, 1999.

Ueno, E.: Beta-1,3-D-glucan, its immune effects and its clinical use. *Jap. J. Soc. Term. Syst. Dis.,* 6: 151-154, 2000.

Verlhac, V., Gabaudan, J., Obaqch, A., Schuep, W., Hole, R.: Influence of dietary glucan and vitamin C on non-specific and specific immune responses if rainbow trout (*Oncorhynchus mykiss). Aquaculture,* 143: 123-133, 1996.

Vetvicka, V.: β-Glucans as immunomodulators. *J. American Nutr. Assoc.,* 3: 31-34, 2001.

Vetvicka, V., Novak, M. *Biology and Chemistry of Beta Glucan - Volume 1: Beta Glucans - Mechanisms of Action,* Bentham Science Publishers, 2011.

Vetvicka, V., Sima, P.: β-Glucan in invertebrates. *Invertebrate Survival J.,* 1: 60-65, 2004.

Vetvicka V., Vetvickova, J.: Immunostimulating properties of two different β-glucans isolated from Maitake mushroom (*Grifola frondosa*). *J. American Nutr. Assoc.*, 8: 33-39, 2005.

Vetvicka V., Vetvickova, J.: An evaluation of the immunological activities of commercially available β1,3-glucans. *J. American Nutr. Assoc.*, 10: 25-31, 2007 a.

Vetvicka V., Vetvickova, J.: Physiological effects of different types of β-glucan. *Biomed. Pap. Med. Fac. Univ. Palacky*, 151: 225-231, 2007 b.

Vetvicka, V., Vetvickova, J.: A comparison of injected and orally administered beta glucans. *J. American Nutr. Assoc.*, 11:42-48, 2008.

Vetvicka, V., Vetvickova, J.: β-Glucan-indomethacin combination and septic shock. *Biomed. Pub.*, 153: 111-116, 2009.

Vetvicka, V., Vetvickova, J.: β1,3-glucan: Silver bullet or hot air? Open *Glycoscience*, 3:1-6, 2010.

Vetvicka, V., Yvin, J.-C.: Effects of marine beta-1,3 glucan on immune reactions. *Int. Immunopharmacol.*, 4: 721-730, 2004.

Vetvicka, V., Baigorri, R., Zamarreno, A. M., Garcia-Mina, J. M., Yvin, J.C.: Glucan and humic acid: Synergistic effects on the immune system. *J. Med. Food*, 13: 863-869, 2010.

Vetvicka, V., Dvorak, B., Vetvickova, J., Richter, J., Krizan, J., Sima, P., Yvin, J.-C.: Orally-administered marine (1-3)-β-D-glucan Phycarine stimualtes both humoral and cellular immunity. Int. *J. Biol. Macromol.*, 40: 291-298, 2007 a.

Vetvicka, V., Terayama, K., Mandeville, R., Brousseau, P., Kournikakis, B., Ostroff, G.: Pilot study: orally-administered yeast β1,3-glucan prophylactically protects against anthrax infection and cancer in mice. *J. American Nutr. Assoc.,* 5: 1-6, 2002.

Vetvicka, V., Thornton, B. P., Ross, G. D.: Soluble β-glucan polysaccharide binding to the lectin site of neutrophil or natural killer cell complement receptor type 3 (CD11b/CD18) generates a primed state of the receptor capable of mediating cytotoxicity of iC3b-opsonized target cells. *J. Clin. Invest.,* 98: 50-61, 1996.

Vetvicka, V., Thornton, B.P., Wieman, T.J., Ross, G.D.: Targeting of NK cells to mammary carcinoma via naturaly occurring tumor cell-bound iC3b and β-glucan-primied CR3 (CD11b/CD18). *J. Immunol.,* 159: 599-605, 1997.

Vetvicka, V., Vashishta, A., Saraswat-Ohri, S., Vetvickova, J.: Immunological effects of yeast- and mushroom-derived β-glucans. *J. Medicinal Food,* 11: 615-622, 2008 a.

Vetvicka, V., Volny, T., Saraswat-Ohri, S., Vashishta, A., Vancikova, Z., Vetvickova, J.: Glucan and resveratrol complex – possible synergistic effects on immune system. *Biomed. Pap. Med. Fac.,* 151: 41-46, 2007 b.

Vetvicka, V., Vetvickova, J., Frank, J., Yvin, J.-C.: Enhancing effects of new biological response modifier β-1,3 glucan sulfate PS3 on immune reactions. *Biomed. Pharmacotherap.,* 62: 283-288, 2008 b.

Vitaglione, P., Sforza, S., Galaverna, G., Ghidini, C., Caporaso, N., Vescovi, P., Fogliano, V., Marchelli, R.: Bioavailability of trans-resveratrol from red wine in humans. *Mol. Nutr. Food Res.,* 49: 495-504, 2005.

Wang, Z., Shao, Y., Guo, Y., Juan, J.: Enhancement of peripheral blood CD8⁺ T cells and classical swine fewer antibodies by dietary β-1,3/1,6-glucan supplementation in weaned piglets. *Transboundary Emerging Dis.,* 55: 369-376, 2007.

Williams, D. L., Di Luzio, N. R.: Glucan-induced modification of murine viral hepatitis. *Science,* 208: 67-69, 1980.

Wojcik, R.: Effect of brewer's yeast *(Saccharomyces cerevisiae)* extract on selected parameters of humoral and cellular immunity in lambs. *Bull. Vet. Inst. Pul.,* 54:181-187, 2010.

Xia, Y., Ross, G. D.: Generation of recombinant fragments of CD11b expressing the functional β-glucan-binding lectin site of CR3 (CD11b/CD18). *J. Immunol.,* 162: 7285-7293, 1999.

Yan, J. : β-Glucan-mediated tumor immunotherapy – mechanisms of action and perspective. In: Vetvicka, V., Novak, M. (Eds.), *Biology and Chemistry of Beta Glucan - Volume 1: Beta Glucans - Mechanisms of Action,* Bentham Science Publishers, 2011, pp. 39-47.

Yoshiba S., Ohno N., Miura S. T., Adachi Y., Yadomae T.: Immunotoxicity of soluble β-glucans induced by indomethacin treatment. *FEMS Immunol. Med. Microbiol., 21,* 171-179, 2001.

Zekovic, D. B., Kwiatkowski, S., Vrvic, M. M., Jakovljevic, D., Moran, C. A.: Neutral and modified (1-3)- β-D-glucans in health promotion and disease alleviation. *Crit. Rev. Biotechnol.,* 25: 205-230, 2005.

Zeman, M., Nosalova, V., Bobvek, P., Zakalova, M., Cerna, S.: Changes of endogenous melatonin and protective effect of diet containing pleuran and extract of black elder in colonic inflammation in rats. *Biologia,* 56: 695-701, 2001.

ACKNOWLEDGEMENTS

The author(s) would like to thank Ms. Cecile Limon and Ms. Rosemary Williams for their valuable assistance in the preparation of this manuscript.